Three Wells

Three Wells

Essays and Lectures on the Biblical, Geometrical, Political Mind

by Robert D. Sacks

RESPONDEO BOOKS
Santa Fe, New Mexico

First published by Respondeo Books in 2022
www.respondeobooks.com

ISBN 978-0-993059-8-0

Respondeo Books
P.O. Box 5648
Santa Fe, NM 87502
United States

Robert Sacks photo courtesy of St. John's College, Santa Fe, New Mexico

Book design by Howard Fisher
Cover design by Adam Robinson for Good Book Developers

Contents

Introduction

The inquiries presented in this volume are diverse. They fall naturally into familiar categories: There is a good deal on the Bible, a book whose richness Robert Sacks has plumbed as very few other readers have done. The sections on political and mathematical themes, too, demonstrate Sacks's attention to texts both in their details and in their broadest and boldest aims. The recollections of Sacks's two most influential teachers, Leo Strauss and Jacob Klein, give the reader a glimpse of how self-knowledge and knowledge of one's world emerge together for one whose life is truly a life of the mind.

There is a kinship of themes among the writings collected here. The conflict between land and sea recurs persistently in the stories of the Israelites' emergence from slavery and reappears in Shakespeare's *Tempest*, in which Prospero almost explicitly takes on the task of reconciling sea and land, of making peace between the chaos of forces both human and non-human and the demands of political life. Sacks alerts the reader to the need for the solid ground defined by Euclid's postulates by pretending to demonstrate the trisection of an angle. But we are reminded that ungrounded claims to understanding may represent a possibility of immediate and original knowing we would not want to dismiss. If there is a lesson to be gleaned from these readings taken together, it comes across as something like this: With no return to the sea, to whatever is at the origin of things, life and thought on the solid dry land becomes sterile and once-seen truths are sedimented into prejudice. But without the order of law and rationality, mere assertion—the more forceful the more successful—and even violence itself, rule. The sea and the land are not simply separable. They mutually define one another, not only in the opening Book of Genesis but in human experience.

The insights of a particular mind demand to be articulated and shared, assessed and incorporated into coherent accounts. Robert Sacks's thoughts present themselves not as definitive statements, and certainly not as lessons to be learned, but as readings of great works. Their authors in turn used the written word to develop a kind of knowledge that can emerge only in response to the partial seeing and persistent questioning of a human being. The mind of the author whose writings are collected here is unique and individual. Sacks's choice to share his thoughts raises a question and perhaps a hope: is the human mind in some sense one? Can an individual, after all, recollect what he has not experienced? Robert Sacks's voice is uniquely his own. He makes intelligible much that even for a diligent reader was obscure. These writings are offered as a sharing of a mind open to other minds, which all derive their being from the same original fluid source of all things.

To what extent can the personal process of recollection of one human being contribute to a similar process for others? Sacks's path of self-discovery begins with his childhood wonder at the fact that another boy's grandfather spoke English rather than Yiddish. It takes him back to the stories of the Bible, a book which at first glance offers the opposite of individual self-inquiry. The Bible seems to indicate a path for recollection in common. The Jewish tradition provides a context in which to live as a member of a people; it responds to the need for a grounding in society shared not merely with one's contemporaries but with the suffering and struggling human beings from whom one has descended and whose stories the Bible conveys. If it risks inducing thoughtless repetition of accepted interpretations of the stories, Sacks's readings check that risk with his observations of the text's subtleties. But this is too simple a description. In Sacks's nuanced account, the reader of the story of David must follow that young man whose journey

> ... seems to require a going back, a knowledge of who he was, where he has come from, and why it was he who had to be king. But having come so far, it seems *we* [my emphasis] are faced with nothing but a knot of contradictions. How can David be a teacher where there is nothing to be taught? How can a lasting tradition be established in the ever-changing sea?" [page 59]

David must respond to constantly shifting circumstances in which he, the anointed king of the Israelites, must run for his life with the sword of Goliath, the Philistine giant, as defense. But the problem is not David's alone. We, the readers, imitate David as we follow his journey. Then, when we read of the generosity of the witch whom Saul prevails upon to revive the ghost of Samuel, Sacks says,

> ... to the extent that he has caused us to fall in love with the witch ... the author has ... caused us all to peer beyond ourselves, to become fish. [page 61]

Sacks has fallen in love with the witch, and he can show us the way to the sea. The reader's education, like David's, takes us outside of and beyond ourselves in pursuit of understanding both of ourselves and of what is utterly beyond us. By contrast with David, our fish-like character endures only for a moment, but for that moment we must come to terms with the challenges that the Biblical account itself presents to dogmatic reading, and to our tendency to rigidify notions of right and wrong.

David's fish-like character is complemented by Job's, whose questioning shakes pious tradition to its roots. Job is the example of a human being who has lost all sense of being at home in the world. In his encounter with God, not in autonomous human activity of any kind, he finds it anew. In Robert Sacks's reading of the Book of Job the characters of each of Job's friends emerge and become comprehensible. Job's inability to dismiss Bildad's arguments subsists alongside his vivid sense that he has not sinned. All Bildad's exhortations cannot induce Job to abandon his sense of the injustice he experiences or to reconcile himself to the inscrutability of God's ways. No response to him is adequate but the direct confrontation with God. What is Job's story to one who has not endured the trials of Job? Whatever else it is, Sacks tells us, it is the example of a human being who has

> a certain light, an innate sense of what is just and what is unjust. For Job, no man can ignore that light as long as he finds it within himself.... [page 77]

Job has found himself compelled to question his understanding and the ways of life he shares with others or else lose all sense of who he is. The divine response may be hard to take, as Elihu warns, for it comes from the world beyond the human horizon within which humans must dwell. Job's confrontation with God permits him to go on in the human world, in which he lives comfortably and acts generously (if not in full accord with convention). Perhaps he can assess things more clearly and on their own terms than he could before. But is each thing what it is by its own nature or does each depend upon God to make it what it is and to preserve it as such? In Job's encounter with God that question is addressed, but in no rational manner. Sacks observes that in "the realm of the jackal," the world beyond man in which God's inscrutable ways are demonstrated,

> the beings . . . may each have their own being, and hence their own strivings and ambitions apart from any human concern; yet man may be open to them. [page 86]

Is the world, in principle at least, intelligible—if not directly to the rational mind then perhaps in some other way, a way represented by the experience of Job? The question arises precisely because the world seems in part intelligible and yet its intelligibility as a whole eludes us: "The world makes too much sense to make no sense, and yet it makes no sense" (pages 77, 83, 159). Since no one can credibly claim to have grasped all things as they are, the question of how they may reveal themselves remains wide open. Perhaps no one has fully grasped any one thing despite the support of a tradition that seems to offer a path to understanding oneself and one's world. Robert Sacks demonstrates that the Biblical tradition contains and embraces the inquiry that arises for a mind alerted to its lack of full understanding. His characteristic evidence of the greatness of the Bible is that it can respond to even the "most impious questions" he can pose (page 180). In doing so, the Bible offers a compelling alternative to the philosophical tradition and demands its respect.

Sacks's readings are not the finished work of an expert whose authority attests that no one else need undertake their own review of these texts, although they represent admirable scholarly inquiry. The Biblical stories, for example, may not have precisely the same resonance for another reader as they have for him. Sacks's work helps this reader at least to find her way into the tradition that both is and is not her own. For me the Biblical stories are in part predecessors to the Christian teaching through which I first heard and read them. That teaching, badly represented as it was in my childhood experience, has contributed, no doubt, to who I am and how I read. There is no defense here of one theological position over another. Sacks's readings of the Bible are the work of one who is as deeply impressed with Athens as with Jerusalem, who is attentive to these complementary traditions that together form the Western world and which provide different responses to the fundamental human problems. When Sacks informs the reader that he studied the Bible with Leo Strauss, he notes that his teacher emphasized the commonality of the theological and philosophical traditions of the west. Both Jerusalem and Athens acknowledge what he calls the problem of the corn and the bread—the distinctions between what is so according to nature and what is the product of human endeavor—and they both insist upon the priority of the love of justice. And yet they diverge: as Strauss puts it, while both traditions pursue wisdom as an end, the philosophical approach that derives from Athens acknowledges no miracles, while the Biblical God is unknowable (pages 115–116). He can therefore be both the origin of the natures of things and the source of miraculous intervention into a world that otherwise appears intelligible. While the theological tradition warns of the inadequacy of reason, the philosophical tradition holds out the possibility, in principle, for humans to understand things as they are and to take their bearings in life from that understanding. That possibility is not, or at least not yet, realized. The Western thinker who cares about justice must look to the Bible as well as to the works of philosophers.

Robert Sacks's thoughts arise in response to the range of great books he addresses here. They are examples of careful, thoughtful reading that are exemplary for any reader who wishes to understand the authors more clearly. They are exemplary because these readings are attempts to see not only how the author has made the text cohere, but also how that very coherence responds to a difficulty, a perplexity or a question: can Shakespeare/Prospero re-found a city by giving his audience an experience of enjoyment of the good? Can Meno's slave (can any of us) not only grasp the truth about the diagonal that makes the double square but continue to see it for himself once it is embedded in an understanding of the general truth about the relations between the sides of a triangle? Once the Euclidean account of that truth has been laid out, how can we revive the immediate recognition of truth that constitutes knowing for oneself?

The problem of sedimentation, the reduction of the knowledge of individual human minds to the prejudices and dogmas, the thoughtless habits of speech and implicit claims that are reinforced constantly by the agreement of others, is both the lasting impediment to and the occasion for learning. Human beings do not come to a sort of Cartesian clearing, Sacks reminds us, and begin anew, at least not with respect to the most fundamental convictions that shape our lives. Instead and at best, we see glimpses of a partial clearing on the occasions when we become aware of a prejudice, a notion that is embedded in our habitual speech and thought that we can recognize as thoughtlessly derived from others. At those moments one can begin to see something clearly. At those moments, too, one is in a position to perceive the thought of another, not as a representative of a doctrine, but as evidence of a mind in a way akin to one's own and engaged with something knowable. For most of us, when the mind is that of the author of a great book, it is almost incomparably better than our own, and yet akin. Sacks brings that kinship a little closer. He demonstrates how to read a great book not as an artifact but in a living exchange, and he engages in this kind of reading repeatedly and with grace.

Robert Sacks's readings of the Bible will shape my future readings—I cannot think of the Israelites' demand for a king "like other peoples'" in the same way as I have done before. I will not be able to read Ptolemy again without entertaining the thought that he seriously considered that the earth moves around the sun. Others will no doubt respond similarly and will be glad to have been informed of what Sacks saw that they had not already seen. But that is not the point of these readings. In these lectures and essays Sacks engages in a kind of dialogue with the authors. He provides an example of what it means genuinely to engage not simply with the texts he treats but with the questions they address. He illustrates how these texts, all great books, remind a serious reader of what he thinks he knows already and show the way to trace each claim back to its origin in questioning. The books are not mere occasions but living representations of the authors who saw something for themselves and so can point the way to others. It is for us to follow the direction in which they point. Impious or just ignorant, the questions any serious reader is likely to pose to these texts can become part of the dialogue. Sacks's writings serve as an intermediary; he is an intercessor who can show us what it means to read mindfully, with a genuine desire to judge aright.

In the Epilogue of this set of essays and lectures, Sacks recollects that for Jasha Klein, "Knowledge ... was not something that could be written down in a book; knowledge demanded our own voice." Robert Sacks's writings are expressions of his voice that in turn allow the voices of the authors of the texts at hand to become more clearly audible. They constitute careful, attentive engagements with the texts of the Bible, Shakespeare's *Tempest,* Plato's *Meno,* Ptolemy's *Almagest* and others. These writings illuminate the texts he engages with and demonstrate the art of careful and thoughtful reading. The works he discusses are great books—books that respond to the questions raised by the reader and which themselves represent the nuanced voices of the authors behind them. Sacks demonstrates what it is to read these works with an openness to the inquiry that lies behind them. At the same time Sacks's voice, a voice that bears witness to a singular, exemplary mind, rings out loud and clear. The reader is welcomed to listen in on a probing dialogue with the authors and interlocutors that have refined and given nuance to Sacks's thinking. In an interview in 1999 Robert Sacks said, "When I try to understand myself, I find that the self I want to understand is partly hidden between dusty books unopened for hundreds of years." These writings constitute a self-examination in which the author, witch-like, brings the hidden self, the ghostly thoughts that once lay buried, to life before the reader's eyes.

Janet Dougherty
St. John's College, Santa Fe, New Mexico
July 2022

Who is a Hebrew?

Before beginning, let me mention that the Hebrew word עֶבֶד (*'eved*) does not distinguish between a slave, a servant, or even a simple worker.

◆

The first to articulate the word *Hebrew* (עִבְרִי, *'ivri*) was the wife of Potiphar, when she called out, "He has brought among us a Hebrew to laugh at us." Later she complained of "the Hebrew servant (עֶבֶד עִבְרִי), whom you have brought among us." Joseph spoke of himself as having been stolen out of "the Land of the Hebrews," although there is no indication that the Hebrews ever had a land. Later we learn that an Egyptian may not eat with a Hebrew. Evidently the word *Hebrew*, or something like it, was part of the Egyptian language prior to Joseph's arrival.

Throughout the middle east the name *Habiru* or *'Apiru* denotes an inferior social class of outlaws, mercenaries, and slaves leading a marginal and sometime lawless existence on the fringes of settled society. These 'Apiru have no common ethnic affiliations and no common language. Potiphar's wife leaves no doubt that, for her, *Hebrew* is a derisive pejorative.

The book of Exodus opens with an account of "the names of the sons of Israel who came to Egypt." From the first, "the children of Israel were fruitful and increased greatly," prompting the king of Egypt to declare that "the people of the children of Israel are too many"; and we learn that "the Egyptians were in dread of the children of Israel". At that point "the Egyptians made the children of Israel work as slaves."

Once slavery was instituted, the word *Hebrew*, never before mentioned in Exodus, would appear seven times before the next occurrence of the term Israel, thirty-two verses later:

> The king of Egypt died, and the children of Israel groaned because of their slavery and cried out for help. Their cry for rescue from slavery rose up to God. And God heard their groaning, and God remembered his covenant with Abraham, with Isaac, and with Jacob. God saw the children of Israel—and God knew. [Exodus 2:23–25]

When the decision to establish slavery was made, the man who had been called *king of Egypt* began to be called *Pharaoh*; but in the verses just quoted it was the *king of Egypt* who died, not *Pharaoh*. Notice also that God heard the groaning of *the children of Israel*, not the groaning of the Hebrews. God then called upon Moses to bring the children of Israel out of Egypt.

Moses realized that the children of Israel would demand to know God's name (Exodus 3:0–12). It is through speech, and through names in particular, that slaves can hold themselves together. It is also the only way they have of connecting the present to the past.

Moses was told:

> Go and gather the Elders of Israel together and say to them, "The Lord, the God of your fathers, the God of Abraham, of Isaac, and of Jacob, has appeared to me, saying, 'I have observed you and what has been done to you in Egypt.' ... And they will listen to your voice, and you and the Elders of Israel shall go to the king of Egypt and say to him, 'The Lord, the God of the Hebrews, has met with us; and now, please let us go a three days' journey into the wilderness, that we may sacrifice to the Lord our God.'"

In this passage we notice that the children of Israel, though slaves, are sufficiently organized to have a body generally recognized as the Elders of Israel. But the Hebrews have no Elders of their own to speak for them.

> Afterwards, Moses and Aaron went and said to Pharaoh, "Thus says the Lord, the God of Israel: 'Let my people go, that they may hold a feast to me in the wilderness.'" But Pharaoh said, "Who is the Lord, that I should obey his voice and let Israel go? I do not know the Lord, and moreover, I will not let Israel go." [Exodus 5:1–2]

Exodus 5:1 marks the only time that Moses speaks to Pharaoh in the name of *the God of Israel*, whereas he speaks to Pharoah in the name of *the God of the Hebrews* six times. From Exodus 5:1 it is clear that Pharaoh knows the name Israel, but it seems clear from his instructions to the midwives (Exodus 1:6) that, left to himself, he prefers the term *Hebrew*. Ironically enough, perhaps the only non-pejorative use of the word is in the phrase "Hebrew slave"—עֶבֶד עִבְרִי (*'eved 'ivri*).

On the other hand, while Exodus speaks seventeen times about bringing the children of Israel out of the land of Egypt, it never says anything about bringing out any Hebrews. And once the Red Sea was crossed, the word "Hebrew" would never again be used to name the Israelites as a whole.[1] This holds true except for the Philistines, to whom we now turn.

◆

The only people other than the Egyptians ever to use the term "Hebrews" were the Philistines. To them, too, it was often a term of contempt, but that contempt did not always redound to their own good repute. On one occasion the Philistine garrison cried, "Look, Hebrews are coming out of the holes where

1 The term "Hebrew" appears again in Exodux 21:2 and Deuteronomy 15:12; but both instances concern the עֶבֶד עִבְרִי, the Hebrew slave, rather than a people.

they have hidden themselves." But it was only one man, Jonathan, accompanied by his armor-bearer, creeping up to cause a panic and take down about twenty men. To understand the Philistines' continual use of the name *Hebrew* we must ask, "Who are the Philistines?" Their name first appears in Genesis 10:13–14, verses generally translated along these lines:

> Egypt fathered Ludim, and Anamim, and Lehabim, and Naphtuhim, and Pathrusim, and Kasluhim, from whom the Philistines came, and Kaftorim.

Not only does this passage feel strange syntactically, it is also at odds with several other verses in scripture. For example:

> Did I not bring up Israel from the land of Egypt, and the Philistines from Kaftor? [Amos 9:7]

> As for the Avvim, who lived in villages as far as Gaza [a Philistine city], the Kaftorim, who came from Kaftor, destroyed them and settled in their place. [Deuteronomy 2:23]

> For the Lord is destroying the Philistines, the remnant of the coastland of Kaftor. Baldness has come upon Gaza; Ashkelon has perished. [Jeremiah 47:5]

These quotations suggest that the words conventionally translated as "from whom the Philistines came" have fallen out of place. Moreover, the word מִשָּׁם (*misham*) in Genesis 10:14 should not be rendered *from whom* but *from where.* So the passage should instead read:

> Egypt fathered Ludim, and Anamim, and Lehabim, and Naftuhim, and Pathrusim, and Kasluhim, and Kaftor, from where the Philistines came.

In the case of every other nation we are told who was its father; and the reader can trace its origin back to Noah. But of the Philistines we are told only that they came from Kaftor. We do not know how they came to be there, nor are we told who "fathered" them. What more is known of Kaftor? Not much.

In the latter days of the reign of King Saul, David was forced to flee, and he spent his time among the Philistines. When he returned as king, he had with him a cadre of men known as כְּרֵתִי (K'reti or Cretans), and פְּלֵתִי (P'leti), which name looks very much like פְּלִשְׁתִּי (Ph'lishti or Philistine). It is not clear whether these were two different peoples, or one people with two names; but in either case it would seem that Kaftor is the island of Crete.

The Philistines appear to be a sea people of unknown origin, lately come from over the western sea together with the giants, who have been on the earth from antediluvian days. Once there came out from the camp of the Philistines a champion named Goliath of Gath, whose height was six cubits and a span.

He was there among the Philistines only because four hundred years earlier, Joshua had attacked the Anakim, driving them out completely from the land of the people of Israel. Only in Gaza, in Gath, and in Ashdod did any remain.[2] Now these Anakim descended from the Nephilim, who "were on the earth in those days, before the flood and also after" (Genesis 6:4). Genesis 6:3 reads:

> My spirit shall not judge from within man forever, for he is flesh; his days shall be 120 years.

The phrase "from within man" denotes the ability of a human to act according to the spirit of the Lord which he finds within. Instead, God's decision to save Noah from the flood meant that mankind would have to live under external law—either the laws given to Noah or the laws given to Moses. If the Philistines, a people of unknown origin who knew the sea, came through the flood with the help of their friends the Nephilim, they would feel themselves exempt from the external laws of Noah. From the Philistine point of view, the Israelites are *Hebrews* because they are bound by an external law.

While the Philistines know the Israelites as the *Hebrews*, Israel, in turn, speaks of the Philistines as the *foreskinned*. To understand this, let us consider Leviticus 19:23–24. A literal rendition of this passage would read:

> When you come into the land and plant any kind of tree for food, then you shall foreskinize its fruit as foreskin. Three years it shall be foreskin to you; it must not be eaten. And in the fourth year all its fruit shall be holy, an offering of praise to the Lord.

The fruits of the fourth year belong to God, just as though they were the first fruits. The world of the giants that was before the flood grew beyond itself to hide its true self; and so we are told:

> Circumcise therefore the foreskin of your heart, and be no longer stubborn. [Deuteronomy 10:16]

It need only be added that it was not Abraham but *Abram the Hebrew* who, in a preview of what was to come, went down into Egypt.[3]

---

2 I Samuel 17:4, Joshua 11:22, Numbers 13:33

3 Genesis 12:10, 14:13

Judah
Seeds of Kingship

One cannot give a reasoned account of the life of Judah without turning first to the birth of his elder brother Reuben, Jacob's firstborn—his בְּכוֹר, (*b'chor*). What a strange being that Hebrew word is. בְּכוֹר can mean the first fruit of a plant, or the firstborn of an animal or of a human being. The strange thing about the word is that it has no more fundamental meaning lying behind it, in contrast to a word like *primogenitor*. It is as primitive a root as the word *bear* (carry), of which *born* is the past participle of the passive voice. The birthright of the firstborn presents itself as a notion as fundamental as the idea of *carrying*. The common availability of the word is what makes Isaac (and initially everyone else) assume that the birthright belonged to Esau.

> If a man has two wives, the one loved and the other unloved, and both the loved and the unloved have born him children, and if the firstborn son (בְּכוֹר) belongs to the unloved, then on the day when he assigns his possessions as an inheritance to his sons, he may not treat the son of the loved as the firstborn (בְּכוֹר) in preference to the son of the unloved, who is the firstborn; but he shall acknowledge the firstborn, the son of the unloved, by giving him a double portion of all that he has, for he is the firstfruits of his strength. The right of the firstborn (מִשְׁפַּט בְּכוֹרה) is his. [Deut. 21:15–17]

The word בְּכוֹר is sometimes confused with the word בָּחוֹר, since the two are indistiguishable in most European dialects; though in most Sepharadic dialects a distinction is still maintained. בָּחוֹר comes from the root בחר, which means *to choose*. The word בָּחוֹר itself means *the chosen one*. In Tanakh the בְּכוֹר, the firstborn, is always the default chosen one. Yet in the most interesting cases things never fall out that way. After the birth of Jacob's eighth son, Asher, Reuben (Jacob's firstborn) found some mandrakes in the field and brought them to his mother Leah. Rachel asked Leah, "Please give me some of your son's mandrakes." Leah agreed, but with the caveat that she, not Rachel, would sleep that night with their husband Jacob.[1] Leah's love for her husband seems to outweigh her desire for the mandrake even though she, too, had ceased to bear children.[2]

Leah's son, Reuben, presented her with what he thought to be a magical fruit with a magical sounding name, דוּדָאִים (*dudaim*). The word seems

1 Gen. 30:14–21

2 Gen. 30:9. The mandrake was believed to be an aphrodisiac and an aid to fertility.

almost to be a dual form of the word used so often in the *Song of Songs* to refer to the beloved.[3] The name of the plant, then, rings as *pair of lovers*.

We have here an exceptionally clear view of Reuben. He is kind, thoughtful and generous; but somehow things never seem to work out quite right for him. The mandrake he so lovingly gave to his mother appears to have been totally useless. Without its aid Leah conceived and bore three more children—two sons and a daughter—while Rachel still had nothing. Reuben was always getting everything mixed up. Shortly after Rachel's death,

> Reuben went and lay with Bilhah, his father's concubine. And Israel heard of it. [Gen. 35:22]

Bilhah had been Rachel's handmaid. With Rachel dead, Reuben rashly assumed that any connection between his father and Bilhah had come to an end.

Nothing is heard again of either Judah or Reuben until one day, when the brothers were out keeping their sheep in Dothan, they saw Joseph coming over the horizon. The brothers, not taking well to Joseph's rather imperious personality, decided to kill that "dreamer of dreams" and throw his body into a pit. Reuben, however was of a kinder nature and, being the eldest, thought it his duty to save Joseph's life. He pleaded with the other brothers, "Shed no blood; throw him into this pit here in the wilderness, but do not lay a hand on him."[4]

But Reuben was also somewhat of a bumbler. His plan was to have the brothers throw Joseph into a pit, thinking to come back later and return the boy to his father. Judah, however, was more thoughtful. He realized that the animosity would only arise again, and that the only solution was to get the boy out of the country. The only way that would be possible was to persuade the brothers to sell Joseph to a passing Ishmaelite caravan, but how was he to convince them to do that?[5]

His argument was somewhat complex. He urged them:

> "What profit is there for us if we kill our brother and conceal his blood? Come, let us sell him to the Ishmaelites, and let not our hand be upon him, for he is our brother, our own flesh." And his brothers listened to him. [Gen. 37:26–27]

While Reuben's plan was more pious, Judah's plan seems to be wiser. This wisdom is displayed in several ways. Judah not only realized that, even if Joseph returned to his father's house, the same problem was sure to arise again; he also realized that Joseph was capable of managing his own affairs even in difficult

---

3 Song 1:13

4 Gen. 37:17–22

5 Gen. 37:22–25

circumstances. Perhaps Judah's greatest wisdom is revealed in the twofold nature of the appeal he made to his brothers. Only after showing them that murdering their brother would bring no monetary gain, did he appeal to their natural abhorrence of fratricide. He spoke both to what was lowest in them and to what was highest. Without the appeal to the lowest, they would never have listened to the highest; and without the appeal to the highest they would have learned nothing. One cannot read the words "and his brothers listened to him" without thinking not only about the present verse but also of the role Judah will play in the later chapters, where the reader might detect signs of David's kingship to come.

When Joseph arrived, the brothers stripped him of his long-sleeved coat, threw him into the pit, and sat down to a meal—apparently out of sight of the pit, thus allowing Reuben to sneak away to free Joseph. That was when they spotted the Ishmaelite caravan and Judah formulated his alternative plan. At this point the text becomes quite unclear. It reads:

> Then Midianite traders passed by. And they drew Joseph up and lifted him out of the pit, and sold him to the Ishmaelites for twenty shekels of silver. They took Joseph to Egypt. [Gen. 37:28]

The problem is the antecedent of the word *they*. Did the brothers see the Midianites coming and run down to pull Joseph out of the pit? Or did they not even notice that there was a caravan of Midianite traders who saw the boy in the pit? In other words, were the Midianites the ones who drew Joseph out of the pit and sold him to the Ishmaelites for twenty shekels?

At any rate, when Reuben returned and announced that the boy was gone, nobody said a word; they simply proceeded as planned. Did they know they were guilty of selling their brother to the Ishmaelites? Or were they left knowing only that the boy was gone, and that they were somehow guilty of some unknown wrong?

When the brothers returned to Jacob they presented him with Joseph's coat steeped in the blood of a goat and said, "This we have found; please to recognize whether it is your son's coat or not." Jacob recognized the coat and could not be comforted.[6]

After what had happened Judah felt he could no longer share a life with his brothers. He no longer placed any hope in his great-grandfather's mission and, in a strangely inverted image of Abraham's journey, he too left his land, his kin, and his father's house to settle down to a new life apart from his brothers. His break with the past seemed as final as Abraham's had been. He

6 Gen. 37:32–36

lodged with a man from Adullam named Hirah; there he met and married a Canaanite woman named Shua, and together they had three sons: Er, Onan, and Shelah.[7]

What Judah did not know is that in leaving his father's house he was already beginning to return there. For David, his far distant descendant, would one day fall from Saul's favor and find no safety among the men of Achish; but the dissidents would rally around him, and David's struggle for the kingship would begin there in Adullam where Judah fled:

> David departed from there and escaped to the cave of Adullam. And when his brothers and all his father's house heard it, they went down there to him.12 [I Sam. 22:1]

David's last great act of battle came very late. The old man was faint, and his soldiers risked their lives to cross through enemy lines in order to fetch their king a single cup of water from a well in Beth-lechem, the town of his birth. One cannot help but be impressed by the dignity with which David poured out the water as a libation rather than drink the cup for which his men risked so much. All this took place where it all began—for David as for Judah—in a place called Adullam, which is never mentioned again in the text.

Judah, then, has left his father's house and has gone off with a friend from the city that holds together the beginning and the end of the active life of his true heir, King David.[8]

As time passed, Judah took a wife for his firstborn son Er; her name was Tamar. But Er died without heir because the Lord judged him wicked and put him to death. It was then up to his brother Onan to raise up a seed for Er; but since the offspring would not be counted his, Onan spilled his semen on the ground. For that act he too was put to death by the Lord. But Tamar still felt it her duty to raise up a seed in memory of those two sons. She thought one day to marry the third son, Shelah, who was then still only a boy; so Judah persuaded Tamar to return to her father until Shelah should grow to manhood. During this time, Judah's wife Shua died.[9]

Time finally came for Shelah's marriage, but Judah, fearing that he might suffer the same fate as his older brothers, never sent for Tamar. But she, having heard one day that Judah would be passing by, threw off her widow's weeds and stood at the city gate dressed as a prostitute, waiting for Judah and Hirah to come by.

---

7 Gen. 38:1–5

8 II Sam. 22:13–17

9 Gen. 38:6–12a

Her plan worked. Judah slept with her and promised to send her a kid as payment. In pledge, she demanded his signet, his bracelet and his staff. But when Judah later sent Hirah to deliver the kid and redeem the pledge, no prostitute was to be found at the city gate. Judah made no fuss over the loss of his pledged items and seemed content to let the affair be forgotten.[10]

Some time later, Judah heard that Tamar was about to give birth by harlotry, and he demanded that she be publicly burnt. But when Tamar appeared, she produced the signet, the bracelet and the staff, and said, "Please to recognize this signet, bracelet, and staff."[11]

Those words, "Please to recognize"—he had heard them before on that terrible day when he and his brothers brought Joseph's blood-drenched coat to their father Jacob. Time suddenly went all ajumble for Judah; was it now, or was it then? Who was it that was speaking? Was Tamar speaking to him, or was it himself speaking to his father Jacob? What was it that he was to recognize? Was it the staff, or was it the coat, or was it something else entirely? He acknowledged his pledges to Tamar and said, "She has been more righteous than I."

Judah had learned from Tamar something about responsibility, and he was now ready to return to his brothers. Judah's return signaled his acknowledgement that the wrong done to Joseph was to be borne by all the brothers, guilty and innocent alike. It is the kind of wrong called *perversion* (עָוֹן, *avon*), as we learn much later when, accused of a theft they did not in fact commit, Judah immediately understands their plight as delayed punishment for the abuse of Joseph long ago:

> Judah said, "What shall we say to my lord? What shall we speak? Or how can we clear ourselves? God has found out the perversion (עָוֹן) of your servants; behold, we are my lord's servants, both we and he also in whose hand the cup has been found." [Gen. 44:16]

Perversion is collective in nature:

> You shall not bow down to them or serve them, for I the Lord your God am a jealous God, visiting the perversion (עָוֹן) of the fathers upon the children, to the third and the fourth generation of those who hate me. [Exod. 20:4]

and is in this respect to be contrasted with sin (חֵטְא) or guilt, which are individual:

> Fathers shall not be put to death because of their children, nor shall

10 Gen. 38:12–23

11 Gen. 38:24–26

> children be put to death because of their fathers. Each one shall be put to death for his own sin (חֵטְא)."[Deut. 24:16]

◆

The text now recounts what happened to Joseph in Egypt: his time spent in the house of Potiphar, his years in prison, and his subsequent rise to power under Pharaoh.

A full generation goes by. When the focus of the story returns to Canaan, we are told that ten of Joseph's brothers went down to buy grain in Egypt.[12] The text stresses the number *ten*. This serves in part to introduce the following verse, which relates that Jacob kept Benjamin back. But the chief significance of emphasizing the number *ten* is to contrast it with the number *nine*; the text is subtly telling us that Judah had returned to his brothers and had found a proper role for himself among them. The reason why the text did not wish to mention Judah explicitly will become clear in what follows.

When the brothers came before Joseph he recognized them but they did not recognize him. He accused them of being spies, but they said:

> "We, your servants, are twelve brothers, the sons of one man in the land of Canaan; and behold, the youngest is this day with our father, and one is no more." [Gen. 42:8–14]

Joseph let them sit two days in custody, then said to them:

> "Do this and you will live, for I fear God: if you are honest men, let one of your brothers remain confined where you are in custody, and let the rest go and carry grain for the famine of your households, and bring your youngest brother to me. So your words will be verified, and you shall not die." And they did so. Then they said to one another, "In truth we are guilty concerning our brother, in that we saw the distress of his soul, when he begged us and we did not listen. That is why this distress has come upon us." [Gen. 42:18–21]

It is strange how in the human mind, the things it knows, the things it does not know, and the things it knows only in a way, interplay and even contradict themselves and each other. The brothers' present thoughts and feelings of guilt seem to have been called forth by Joseph's indication that he feared God. Joseph spoke to them through an interpreter, yet the words recorded could be taken either as "I fear the gods" or "I fear God." Perhaps it was just the timbre of his voice, or his insistence that they share some common guilt; but something about this man recalled Joseph to their minds, and yet at the same time they speak freely of their guilt, as if Joseph spoke no Hebrew. This otherwise

12 Gen. 42:3

insignificant moment is a clear example of how the text views forgetting and remembering, seeing and not seeing. This view is characteristic of the book as a whole and the significance of *tradition* is based on it.

Responding to his brothers' owning up to their responsibility, Reuben burst out:

> "Did I not tell you not to sin against the boy? But you did not listen. So now there comes a reckoning for his blood." [Gen. 42:22]

What Reuben said was perfectly true, but coming as it did, after the brothers' heartfelt acknowledgement, his speech simply fell flat. Like most instances of "I told you so," it served no purpose. Judah, however, for the moment remained silent.

When the brothers returned to their father and told him what *the man* had demanded, Jacob only said to them:

> "You have bereaved me of my children. Joseph is no more, and Simeon is no more, and now you would take Benjamin." But Reuben said to his father, "Kill my two sons if I do not bring him back to you. Put him in my hands, and I will bring him back to you." [Gen. 42:36–37]

Well-meaning Reuben has blundered again. His suggestion arises from his true desire to accept the responsibilities laid upon the firstborn. But Jacob's whole life—his separation from Esau,[13] his diffidence at Shechem[14]—had been spent in the attempt to avoid such sacrifices. Mention of Joseph brings up in Jacob thoughts of a bloodstained coat and the animal that may have caused it. What kind of a father has he been? As he reflects on the losses he has already suffered, Reuben's offer of further sacrifice could only have sounded grotesque.

More time passes, and the famine becomes worse; but Judah still waits in silence. The next chapter begins:

> Now the famine was severe in the land. When they had eaten the grain that they had brought from Egypt, their father said to them, "Go again, buy us a little food." [Gen. 43:1–2]

"A little food." We can picture Jacob, imagining himself in *the man's* presence, begging for a little food. The word *little* stresses the humility of his request and at the same time reveals the complex way in which the human mind works. If *the man* were present, Jacob's humility might persuade him to grant the food, without demanding that Benjamin be brought to Egypt. The figure of *the man* so impresses itself upon Jacob's mind that he acts as if the man were present, even though Jacob knows that he is far away and that, in fact, Benjamin must

13 Gen. 33:1–16

14 Gen. 34:30

be sent. Jacob is in such distress that his mind breaks in two—one half living in the world which is, and the other half living in the world which should have been.

> But Judah said to him; "The man swore to us and said 'You will not see my face unless your brother is with you.' So if you want to send our brother with us, we will go down to buy food; but if you do not want to send him we will not go down, because the man said, 'You will not see my face unless your brother is with you.'" [Gen. 43:3–5]

Judah has finally begun to speak. His words are simple, straightforward and unmistakably clear. He shows almost the same kind of stark coldness that characterized Abraham's three-day journey with Isaac and the stack of firewood. Unlike Reuben, Judah does not begin with a great oath, but calmly describes the situation as best he can.

Israel's answer was:

> "Why did you treat me so badly as to tell the man that you had another brother?" They replied, "The man questioned us carefully about ourselves and our kin, saying, 'Is your father still alive? Do you have another brother?' What we told him was in answer to these questions. Could we in any way know that he would say, 'Bring your brother down'?" [Gen. 43:6–7]

Judah does not answer his father's question but, falling silent, allows his brothers to do so. Their answer is, of course, a lie, at least in a superficial sense; they themselves were the first to mention their father and their two other brothers. But in a deeper sense their words will turn out to be true. Joseph will use almost the same words they have attributed to him when he asks about the welfare of their father.

They may be telling the truth in an even more profound way. We have seen that the seeds of recognition are buried deep inside the brothers. Perhaps it really does seem to them as if Joseph had in fact asked those questions which must surely have been on his mind.

> Then Judah said to Israel his father, "Send the boy with me, and we will arise and go, so that we may live and not die, both we and you and also our little ones." [Gen. 43:8; see also 42:2]

Judah ignores the argument and returns to the simple facts at hand. But he does try to shift the grounds of the discussion a bit. The words "so that we may live and not die," which he amplified by saying "both we and you and also our little ones" are a direct quotation from Jacob, who had used them when he sent the brothers to Egypt the first time. In doing so, Judah is trying to annul Jacob's humility and place the discussion back on the simple and forthright level it had in the beginning, when Jacob first sent his sons for food without ceremony. Judah went on to declare,

> I will be a pledge of his safety. From my hand you will require him. If I do not bring him back to you and set him before you, then let me bear the blame forever. [Gen. 43:8–9]

Judah's words are partly directed to Jacob and partly to himself. Insofar as they are directed to Jacob, they are intended as a correction to Reuben's rather clumsy attempt to persuade their father let them take Benjamin down into Egypt. Judah realized that pledging his own life was no solution, and that the pledge of his own honor would be of greater value to his father. To this extent his words are also addressed to Reuben. Judah and Reuben have one thing in common; they both refused to commit fratricide. But in spite of his refusal, Reuben, the elder, proved to be inadequate as a leader. Judah, in his decision to return to his brothers, knew that he would have to take on the responsibility of the firstborn. Perhaps Judah also senses, as the biblical author knows, that this responsibility will continue even past the days of Josiah.

Reuben and Judah were the two brothers who refused to commit fratricide. But Judah, by taking the place of his brother Reuben, committed fratricide in a deeper sense. And as we shall see, Judah's tribe will live apart from the others, beyond the border of what had been the promised land. In that sense Judah ironically turns out to be the only one of the brothers to have committed a fratricidal act.

To the extent that Judah's words are directed to himself, his private thoughts no doubt go back to the time he spent with Tamar. And now he pledges himself to become the replacement for the bracelet, staff and signet ring he had given Tamar in pledge.

Judah then added, "If we had not delayed, we would now have returned twice,"[15] subtly implying that they should have returned to Egypt immediately in order to save Simeon, as Reuben had urged at the time. But at that time, Jacob had been incapable of hearing it. What Reuben did not see was that speech is of no avail, in fact it can barely be called speech, when it cannot be heard. This insight on the part of Judah helps to explain his periods of silence.

◆

The brothers arrive in Egypt and are invited, or perhaps summoned, to appear at the house of *the great man* at noon. It was all so confusing. Were they being accused of having stolen the money that was later found in their

---

15 Gen. 43:10

bags? No, it could not be that; the man at the gate said, "Peace to you, do not be afraid. Your God and the God of your father has put treasure in your sacks for you. I received your money."[16] But how did he know that? We readers, of course, know that it was not God but the very man with whom they were speaking who placed the money in the sacks, at Joseph's behest.

When Joseph arrived, the sight of his brother Benjamin moved him to tears, but he was able to gain control over himself and ordered the food to be served. The meal was somewhat awkward. Joseph's servants sat apart from him because an Egyptian may not eat with a Hebrew; and Joseph sat apart from his brothers because an Egyptian may not eat with Hebrews. Joseph is both a Hebrew and an Egyptian, and he is also neither.[17]

The brothers were mysteriously seated at the table according to their ages and were fed from the great man's table. Benjamin was given five times as much as the others. All rejoiced and none complained.[18] Once more Joseph had his steward secretly place their money back into their sacks; but this time he had his chalice placed in Benjamin's sack. The brothers left, but after they had gone a short distance Joseph sent the steward after them to accuse them of having stolen the chalice. He was directed to say,

> "Why have you repaid evil for good? Is it not from this that my lord drinks, and by this that he practices divination? You have done evil in doing this thing." [Gen. 44:1–5]

When the steward caught up with the brothers, he added,

> "He who is found with it shall be my servant, but the rest of you shall be blameless." [Gen. 44:10]

To increase the suspense, the steward began his search with the eldest. But when the cup was found in the sack of their youngest brother they all returned to the city together with him. When they stood in front of the great man, Joseph said,

> "What deed is this that you have done? Do you not know that a man like me can indeed practice divination?" But Judah answered, "What shall we say to my lord? What shall we speak? Or how can we clear ourselves? God has found out the perversion of your servants; behold, we are my lord's servants, both we and he also in whose hand the chalice has been found." [Gen. 44:15–16]

Judah has finally emerged as spokesman for the brothers. The simplicity of his speech is in sharp contrast to Joseph's magic. He disdains any attempt to

---

16 Gen. 43:16–23

17 Gen. 43:32

18 Gen. 43:33–34

discover which brother was guilty. Ever since his return from Timnah it had been clear to him that each brother must be responsible for the others. His insight was that unity is even more important than discovering whether their sufferings are just or not.

But Joseph insisted:

> "Far be it from me that I should do so! Only the man in whose hand the chalice was found shall be my servant. But as for you, go up in peace to your father." Then Judah came near to him and said, "Oh, my lord, please let your servant speak a word in my lord's ears, and do not let your anger burn against your servant, for you are like Pharaoh himself." [Gen. 44:17–18]

Judah speaks with the great man in private . He addresses him as "my lord" and treats him with all respect, but the very fact of privacy begins to place them on the same level.

This was Judah's tale:

> When we went back to your servant my father, we told him the words of my lord. And when our father said, "Go again, buy us a little food," we said, "We cannot go down. If our youngest brother goes with us, then we will go down. For we cannot see the man's face unless our youngest brother is with us." Then your servant my father said to us, "You know that my wife bore me two sons. One left me, and I said, 'Surely he has been torn to pieces,' and I have never seen him since. If you take this one from me also and any harm comes to him, you will bring down my gray hairs in evil to Sheol" ... How can I go back to my father if the boy is not with me? I fear to see the evil that would find my father. [Gen. 44:24–34]

Judah simply and calmly presents the situation to Joseph, just as he had to Jacob. His account is a fairly accurate statement of what Joseph might have asked, and paints a clear picture of the situation as it was in Canaan. Judah seems to understand Joseph's desires; but the rest of his speech is, in part, an accusation. How can Joseph hear Judah's words, "our father," without knowing that it was his father as well? Joseph's magic has come close to causing his own father's death. Judah's courage in speaking up and holding the man of power responsible for his own actions is most remarkable. After Judah's implicit allegation,

> Joseph could not control himself before all those who stood by him. He cried, "Make everyone go out from me." So no one stayed with him when Joseph made himself known to his brothers. [Gen. 45:1–2]

Judah's speech was more effective than the sight of Benjamin, and Joseph was unable to restrain himself any longer. The word which has been translated *made himself known* will appear only one more time in the Bible. It is the word which God used to describe His own actions with regard to prophets.

> If there is a prophet among you, I the Lord make myself known to him in a vision; I speak with him in a dream. [Num. 6:12]

As we shall see, it is not accidental that the text uses such an imperious word. For the moment, though, our attention must go from Reuben to Joseph.

After the family had been reconciled and Jacob brought down to Egypt, word came to Joseph that his father, by then a blind old man, had fallen ill. He took his two young sons, Manasseh and Ephraim, to be blessed.

> When Israel saw Joseph's sons, he asked, "Who are these?" Joseph said to his father, "They are my sons, whom God has given me here." And he said, "Bring them to me, please, that I may bless them." ... Joseph took them both, Ephraim in his right hand toward Israel's left hand, and Manasseh in his left hand toward Israel's right hand, and brought them near him. And Israel stretched out his right hand and laid it on the head of Ephraim, who was the younger, and his left hand on the head of Manasseh, crossing his hands, for Manasseh was the firstborn. [Gen. 48:1–15]

Israel's decision to reverse the order of his grandsons had apparently been made even before he met them, for he had asked, "Who are these?" His decision to place the younger before the elder, therefore, could not have been made on the basis of merit. This raises the question: Who is the בָּחוֹר, the chosen one? Is it the best, or is it the בְּכוֹר, the firstborn? The one fosters a solid foundation; the other, continued stability.

One cannot help but compare Jacob's last days with those of his father Isaac. Two blind old men, each about to die; both have blessed the younger over the elder. Each has done his part: Jacob wittingly, Isaac unknowingly guided by the hand of Rebecca his wife, who had somehow seen that the time of the firstborn had not yet come. But Joseph, that great magician who nourished his brothers in Egypt and was praised by all, assumed that the blessing would go to his eldest son; for that is the way things work in a smooth and well-running society, which he himself had helped to establish. Joseph believed himself to have completed the work of the founders. He assumed that from that point on, nothing was needed except to follow the way that had been set. But Jacob was wiser and knew that permanence had not yet been achieved.

Beth-lechem, the burial place of Joseph's mother Rachel, is often referred to by its ancient name, Ephrath. In Hebrew the words *Ephrath* (אֶפְרָת) and *Ephraim* (אֶפְרַיִם) are etymologically related, and their generic forms turn out to be identical. In other words, the Hebrew word for a man from Ephraim is identical to the Hebrew word for a man from Ephrath. They are both called

Ephrati (אֶפְרָתִי); although some English translations distinguish between an Ephratite and an Ephraimite.

The city of Beth-lechem is referred to as Ephrath in six biblical passages. In Genesis it is connected three times with the death of Rachel;[19] and since it was the burial place of Ephraim's grandmother, and virtually bore his name, one would have expected the city of Beth-lechem to have fallen to the lot of Ephraim. In fact Beth-lechem lay some distance to the southeast of Ephraim, within the borders of Judah.[20]

So far as I have been able to establish, the only characters from Beth-lechem who were ever called Ephrati were Elimelech the husband of Naomi, his three sons, and his most illustrious decendent, King David.[21] The point of all this confusing similarity of names is that Joseph believes himself to be the last of the great founders. The future leaders would then naturally be Ephrati from the tribe of Ephraim. Ironically, the leaders will indeed be Ephrati, but instead of being the descendants of Ephraim, they will be the descendants of Judah from the city of Ephrath.

To make the irony even more bitterly complex, Joseph was right in a way he could not have predicted and would not have wanted. Jeroboam, the first king of the northern kingdom of Israel—whose sin, as we are reminded twenty times, kept the kingdom divided—was the son of Nebat, an Ephraimite.[22]

At the end of his life Jacob had a word for each of his sons. Of Reuben he said:

> Reuben, you are my firstborn, my might, and the firstfruits of my strength, preeminent in dignity and preeminent in power. Unstable as water, you will not have preeminence, because you went up to your father's bed and defiled it—he went up to my couch! [Gen. 49:3–4]

Let us review the life of Reuben. He was Jacob's firstborn. In the early days, whenever the brothers were all mentioned together, his name always appeared first. That was true when they met their uncle Esau[23] and when Jacob's sons were first enumerated in Genesis 46:8. That order is continued in later books. At the very beginning of Exodus a list will be given of "the souls that came out of Canaan with Jacob," and again Reuben's name will appear first. The Book of Numbers begins in the same way;[24] but when the tribes are lined up for

19 Gen. 35:16, 19; 47:7

20 Judg. 17:7

21 Ruth 4:11, I Sam. 17:13

22 I Kings 11:26

23 Gen. 33:1–16

24 Num. 1:5

marching through the desert only one chapter later, Judah's name will suddenly emerge at the head of the list.[25]

With his "I told you so" rebuke (Gen. 42:22), we also saw Reuben begin to pull away from his brothers, as indeed Judah had done before him; but, unlike his younger brother, Reuben will never be able fully to return. The tribe of Reuben confirmed their forefather's withdrawal by choosing to live away from the others in the conquered land on the other side of the Jordan.[26]

At the end of the journey to the promised land, when the tribes were about to cross over the Jordan, Moses charged them to stand upon the two great mountains that overlook the city of Shechem: on Mount Gerizim to pronounce blessings, and on Mount Ebal to pronounce curses. Here, where one might have expected to find the names of Leah's first four sons in their normal order—Reuben, Simeon, Levi, and Judah—the verse reads instead:

> When you have crossed over the Jordan, these will stand on Mount Gerizim to bless the people: Simeon, Levi, Judah, Issachar, Joseph, and Benjamin. And these will stand on Mount Ebal for the curse: Reuben, Gad, Asher, Zebulun, Dan, and Naphtaly. [Deut. 27:12]

Issachar is Leah's late-born son, born after she had ceased to bear, after she had twice turned to Zilpah. Issachar has replaced Reuben and Reuben is again first on the list—first, however, on the list of those that stand on the mountain of curses. As for Moses's final blessing, the English Standard Version reads "Let Reuben live, and not die, but let his men be few." The King James Version, in contrast, has "Let Reuben live, and not die; and let not his men be few." A more literal translation of this verse would read:

> Let Reuben live, and not die, and let his men be a number. [Deut. 33:6]

It is probably best to let that ambiguity stand. Jacob's words to his son Judah were these:

> Judah, your brothers will praise you; your hand will be on the neck of your enemies; your father's sons will bow down before you. [Gen. 49:8]

Jacob is primarily foretelling the kingship, which the House of Judah will one day hold; but certainly he must also be thinking back to the dreams which Joseph had dreamt those many years before. Joseph, the great interpreter of dreams, seems to have misinterpreted his own dream. The sheaf which stood in the center while the other sheaves bowed down before it was not his own; it was Judah's.

25 Num. 2:3

26 Num. 32:5

Moses

Father of Many, Father of None

PROLOGUE

Near the end of his first encounter with God, Moses said, "O my Lord, I am not a man of words, either in the past or since you have spoken to your servant, but I am heavy of speech and heavy of tongue" (Exod. 4:10). What did God want of this shepherd who once, through his passion for justice, had slain an Egyptian?

> And it came to pass in those days, when Moses was grown, that he went out to his brothers and looked on their burdens; and he spied an Egyptian beating a Hebrew, one of his brothers. And he looked this way and that way, and when he saw that there was no man, he struck (וַיַּךְ) the Egyptian and hid him in the sand. [Exod. 2:11–12]

The reader cannot understand Moses' claim to be heavy of tongue. In fact Moses seems to be a rather gifted speaker; but, as we shall see, God understood and in anger sent for Aaron.

> Then the anger of the Lord was kindled against Moses and he said, "Is there not Aaron, your brother, the Levite?" [Exod. 4:14]

Most readers suppose that Moses' "heaviness of tongue" means either that he had a speech defect or that he spoke little Hebrew (being more at home in Egyptian)—and therefore that the fine Hebrew we find on every page of Exodus is Aaron's revision or translation of Moses' words. In neither case however, can one see anything to provoke God's anger. Why, then, was God so angry; and what kind of man is this Aaron?

When Moses came down from the mountain with the first set of laws, (Exod. 20–23) the ones suggested by Jethro (Exod. 18:17–24), it was Aaron who was most aware of their insufficiency. He saw that the laws dealt almost exclusively with the daily affairs of mankind and left unaddressed other human needs—in particular, the need for something that would lead the people beyond themselves. The pagan world seemed to them to promise such elevation:

> And all the people took of the golden earrings which were in their ears and brought them to Aaron. And he received them from their hand and fashioned with a graving tool a molten calf; and they said, "These are your gods, O Israel, which brought you up out of the land of Egypt." [Exod. 32:3–4]

But Aaron had no solution other than to erect an altar in the face of the idol and proclaim a feast to the Lord:

> And when Aaron saw it, he built an altar before it; and Aaron made a proclamation and said, "Tomorrow is a feast to the Lord." [Exod. 32:5]

PART ONE

Our story begins when Moses and his people were about to set out from the Wilderness of Sin. That is where the people of Israel had complained against Moses and Aaron (Exod. 16:2), lamenting the loss of the meat pots of Egypt and threatening to kill both of them. God heard their "grumbling" and told Moses that he would give them quail at twilight. At the same time he established a more permanent solution by giving the commandment concerning the morning collection of Manna, a strange kind of food that was to be their main source of sustenance during their journey through the wilderness (Exod. 16:13–19).

> Moses said to Aaron, "Say to the whole congregation of the people of Israel, 'Come near before the Lord, for he has heard your grumbling.'" And as soon as Aaron spoke to the whole congregation of the people of Israel, they looked toward the wilderness, and behold, the glory of the Lord appeared in the cloud. [Exod. 16:9–10]

Here, as throughout Exodus 16, we can see that Aaron had a greater feeling of closeness to the congregation. He had grown up with them, whereas Moses had been a stranger to whom they once could even have said, "Who made you a prince and a judge over us?" (Exod. 2:14).

Both the grumbling and the gift of Manna took place in the Wilderness of Sin, an area in the southwestern part of the Sinai peninsula, just north of Mount Sinai itself. But as the people moved south they found that there was no water to drink. Then the people quarreled with Moses and said, "Give us water to drink."

> So Moses cried to the Lord, "What shall I do with this people? They are almost ready to stone me." And the Lord said to Moses, "...take in your hand the staff with which you struck the Nile, and go..., and you shall strike (וְהִכִּיתָ) the rock, and water shall come out of it, and the people will drink." And Moses did so in the sight of the elders of Israel. And he called the name of the place Massah and Meribah, because of the quarreling of the people of Israel, and because they tested the Lord by saying, "Is the Lord among us or not?" [Exod. 17:1–7]

Time passed, and they reached Sinai, where the law was given and the molten calf worshiped.

> Then in the second year, in the second month, on the twentieth day of the month, the cloud lifted from over the tabernacle of the testimony, and the people of Israel set out by stages from the wilderness of Sinai. [Num. 10:11–12]

From Sinai they headed north. On their way, Moses set up camp in the Wilderness of Paran (Num. 12:16), where God told him to send out one man from each tribe to scout out the land they were to inherit. The scouts were to

> "See what the land is, and whether the people who dwell in it are strong or weak, whether they are few or many, and whether the land that they dwell in is good or bad, and whether the cities that they dwell in are camps or strongholds, and whether the land is rich or poor, and whether there are trees in it or not. Be of good courage and bring some of the fruit of the land." Now the time was the season of the first ripe grapes. [Num. 13:18–20]

The scouts reported:

> "We came to the land to which you sent us. It flows with milk and honey, and this is its fruit. However, the people who dwell in the land are strong, and the cities are fortified and very large...."
>
> But Caleb quieted the people before Moses and said, "Let us go up at once and occupy it, for we are well able to overcome it." Then the men who had gone up with him said, "We are not able to go up against the people, for they are stronger than we are." So they brought to the people of Israel a bad report.... "[It] is a land that devours its inhabitants.... And there we saw the Nephilim, the sons of Anak, who come from the Nephilim, and we seemed to ourselves like grasshoppers, and so we seemed to them." [Num. 13:27–33]

The congregation became disheartened and cried, "Let us choose a leader and go back to Egypt" (Num. 4:14). That was the beginning of the rebellion that led God to declare:

> None of these men, who have seen my glory and my signs that I did in Egypt and in the wilderness, and yet have put me to the test these ten times and have not obeyed my voice, shall see the land that I swore to give to their fathers. And none of those who despised me shall see it. But my servant Caleb, because he has a different spirit and has followed me fully, I will bring into the land into which he went, and his descendants shall possess it. [Num. 14:22–24]

God spelled out his meaning this way:

> Your children shall be shepherds in the wilderness forty years and shall suffer for your faithlessness, until the last of your dead bodies lies in the wilderness. According to the number of the days in which you spied out the land, forty days, a year for each day, you shall bear your iniquity forty years, and you shall know my displeasure. [Num. 14:33–34]

When the people left Mount Sinai they headed north, assuming that they would soon be entering the promised land from the south. As they know now, however, that will not happen. Yet on they go, on to an unknown future. Many years before, long before the time of slavery in Egypt, Esau had invited his brother Jacob to join him along the way. Jacob then answered:

> My Lord knows that the children are frail, and that the nursing flocks and herds are a care to me. If they are driven hard for one day, all the flocks will die. Let my Lord pass on ahead of his servant, and I will lead on slowly, at the pace of the livestock that are ahead of me and at the pace of the children, until I come to my Lord in Seir. [Gen. 33:12–14]

Was this the moment, here in the Wilderness of Paran, that Jacob had in mind—the moment of their need? Perhaps Moses thought Esau's invitation might still be good. He may even have heard of an old tradition, according to which God had once said:

> "To Isaac I gave Jacob and Esau. And I gave Esau the hill country of Seir to possess, but Jacob and his children went down to Egypt," [Josh. 24:4],

and so Moses may even have believed that God himself had prepared the way.

In the course of their travel northwards, the entire congregation of the people of Israel

> came into the Wilderness of Zin... [but] there was no water for the congregation. And they assembled themselves together against Moses and against Aaron. "Why have you brought the assembly of the Lord into this wilderness, that we should die here...? And why have you made us come up out of Egypt to bring us to this evil place? It is no place for grain or figs or vines or pomegranates, and there is no water to drink." Then Moses and Aaron went from the presence of the assembly to the entrance of the tent of meeting and fell on their faces. [Num. 20:1–7]

We have seen this before: previously in the Wilderness of Sin, this time in the Wilderness of Zin, further north. The name has changed little, the people even less. Despite all they had been through together, the people were just as rowdy, just as rebellious. Nothing had changed. And nothing would change until all things changed.

> The Lord spoke to Moses, saying, "Take the staff, and assemble the congregation.... and speak to the rock...you shall bring water out of the rock for them [to drink]. And Moses took the staff from before the Lord, as he commanded him. Then ... Moses lifted up his hand and struck (וַיַּךְ) the rock with his staff twice, and water came out abundantly, and the congregation drank, and their livestock. But the Lord said to Moses and Aaron, "Because you did not believe in me, to uphold me as holy in the eyes of the people of Israel, therefore you shall not bring this assembly into the land that I have given them." These are the waters of Meribah,

> where the people of Israel quarreled with the Lord, and through them he showed himself holy. [Num. 20:8–13]

Now we can begin to see what God had clearly understood so long ago: Moses's heavy lips and heavy tongue. Like Melville's Billy Budd, Moses—who once described himself as *not being a man of words*—when faced with the gravest of transgressions lost all power of speech. He was commanded to speak but his tongue grew heavy. It would not move. He could only lash out, striking the rock twice, as he had previously struck the rock and struck the Egyptian taskmaster. [1] Moses knew from the very beginning that a day would come when, unable to speak, he would strike out. This man who struck an Egyptian and hid his body in the sand was doomed never to enter the promised land. Perhaps God knew that only one possessed of such righteous anger could be a founder. But he also knew that such a founder has not the kind of soul that can continue to live among his people as one of them. Perhaps God was angry, not because of Moses' act, but because no human soul is capable of being both the soul of Moses and the soul of Aaron.

When they arrived at the border, Moses sent messengers from Kadesh to the king of Edom:

> "Thus says your brother Israel: ... The Egyptians dealt harshly with us and our fathers.... And here we are in Kadesh, a city on the edge of your territory. Please let us pass through your land. We will not pass through field or vineyard, or drink water from a well. We will go along the King's Highway. We will not turn aside to the right hand or to the left until we have passed through your territory." But Edom said to him, "You shall not pass through, lest I come out with the sword against you." [Num. 20:14–21]

That marked the end of the southern campaign. They would be forced to go around Edom and continue their journey northwards by finding their way east of the Jordan. Then one day they would cross over the Jordan, going westward to attack and conquer the promised land from the north.

But the people became impatient on the way.

> And the people spoke against God and against Moses, "Why have you brought us up out of Egypt to die in the wilderness? For there is no food and no water, and we loathe this worthless food." Then the Lord sent fiery serpents (הַנְּחָשִׁים הַשְּׂרָפִים, *han'khashim has'raphim*) among the people, and they bit the people, so that many people of Israel died. And the people came to Moses and said, "We have sinned, for we have spoken against the Lord and against you. Pray to the Lord, that he take

1 The Hebrew verb used in connection with Moses' striking the rock is the same as that used when he struck the Egyptian.

> away the serpents from us." So Moses prayed for the people. And the Lord said to Moses, "Make a fiery-one (שָׂרָף, *saraph*) and set it on a pole, and everyone who is bitten (כָּל־הַנָּשׁוּךְ, *kol hanashukh*), when he sees it, shall live." So Moses made a bronze serpent (נְחַשׁ נְחֹשֶׁת, *n'khash n'khoshet*) and set it on a pole. And if a serpent bit anyone, he would look at the bronze serpent and live. [Num. 21:5–9]

Note the continuing Hebrew wordplay throughout this passage. Ordinarily, quieting the people would have been Aaron's task; but by then Aaron was dead (Num. 20:22–29), and Moses was left to perform Aaron's work. But Moses was not Aaron; and a bronze serpent is not a "fiery-one." At first all seemed to go well; but five hundred years later Hezekiah, King of Judah,

> removed the high places and broke the pillars and cut down the Asherah. And he broke in pieces the bronze serpent that Moses had made, for until those days the people of Israel had made offerings to it. It was called Nehushtan (Bronze One). [II Kings 18:4]

PART TWO

In their second encounter, God and Moses had the following interchange:

> Then the Lord said to Moses, "Go in, tell Pharaoh king of Egypt to let the people of Israel go out of his land." But Moses said to the Lord, "Behold, the people of Israel have not listened to me. How then shall Pharaoh listen to me, for I am of uncircumcised lips?" But the Lord spoke to Moses and Aaron ...: to bring the people of Israel out of the land of Egypt. [6:10–13]

The following chapter continues the story with this verse:

> The Lord said to Moses, "See, I have made you like God to Pharaoh, and your brother Aaron shall be your prophet. [Exod. 7:1]

But there is a curious intervening passage which begins:

> These are the heads of their fathers' houses. The sons of Reuben, the firstborn of Israel: Hanoch, Pallu, Hezron, and Carmi; these are the clans of Reuben. The sons of Simeon: Jemuel, Jamin, Ohad, Jachin, Zohar, and Shaul, the son of a Canaanite woman; these are the clans of Simeon. [Exod. 6:14–15]

Since the passage begins by giving a full account of the sons of Jacob's first two sons, one might expect a similar account of the sons of all the others: Levi, Judah, Zebulun, etc. Instead, we get the following detailed account of the sons of Levi alone:

> These are the names of the sons of Levi according to their generations: Gershon, Kohath, and Merari, the years of the life of Levi being 137 years. The sons of Gershon: Libni and Shimei, by their clans. The sons

> of Kohath: Amram, Izhar, Hebron, and Uzziel, the years of the life of Kohath being 133 years. The sons of Merari: Mahli and Mushi. These are the clans of the Levites according to their generations. Amram took as his wife Jochebed his father's sister, and she bore him Aaron and Moses, the years of the life of Amram being 137 years. The sons of Izhar: Korah, Nepheg, and Zichri. The sons of Uzziel: Mishael, Elzaphan, and Sithri. Aaron took as his wife Elisheba, the daughter of Amminadab and the sister of Nahshon, and she bore him Nadab, Abihu, Eleazar, and Ithamar. The sons of Korah: Assir, Elkanah, and Abiasaph; these are the clans of the Korahites. Eleazar, Aaron's son, took as his wife one of the daughters of Putiel, and she bore him Phinehas. These are the heads of the fathers' houses of the Levites by their clans. [Exod. 6:16–25]

Note that although we are told the names of Aaron's wife, his children, and grandchildren and even the names of the sons of Korah, nothing at all is said of either Moses' wife or his children; the text is totally silent. We must remember that.

Next we are assured, curiously enough, that the Aaron and Moses given in this lineage are indeed the same Aaron and Moses who went to speak to Pharaoh:

> These are the Aaron and Moses to whom the Lord said: "Bring out the people of Israel from the land of Egypt by their hosts." It was they who spoke to Pharaoh king of Egypt about bringing out the people of Israel from Egypt, this Moses and this Aaron. [Exod. 6:26–27]

The final phrases of this episode are by far the strangest. They read:

> On the day when the Lord spoke to Moses in the land of Egypt, the Lord said to Moses, "I am the Lord; tell Pharaoh king of Egypt all that I say to you." But Moses said to the Lord, "Behold, I am of uncircumcised (עֲרַל) lips. How will Pharaoh listen to me?" [Exod. 6:28–30]

In these verses the text seems to be addressing the reader directly, telling him what Moses had once said on a certain day in the past. Yet there was no need to do so, since the reader was virtually present at the time and heard it for himself. All these oddities underline the importance of the verse.

What does it mean to have uncircumcised (literally, "foreskinned") lips? The significance of circumcision in general can best be seen by looking at the ways it appears in metaphorical contexts:

> Circumcise (וּמַלְתֶּם) therefore the foreskin (עָרְלַת) of your heart, and be no longer stubborn. [Deut. 10:16]

> When you come into the land and plant any kind of tree for food, then you shall regard its fruit as forbidden. Three years it shall be forbidden to you; it must not be eaten. And in the fourth year all its fruit shall be holy, an offering of praise to the Lord. But in the fifth year you may

> eat of its fruit, to increase its yield for you: I am the Lord your God. [Lev. 19:23–25]

Or, more literally.

> When you come into the land and plant any kind of tree for food, then you shall foreskinize (וַעֲרַלְתֶּם) its fruit as its foreskin (עָרְלָתוֹ). Three years it shall be foreskin to you; it must not be eaten. And in the fourth year all its fruit shall be holy, an offering of praise to the Lord. But in the fifth year you may eat of its fruit, to increase its yield for you: I am the Lord your God.

The piñon trees around my house yield their nuts about every seven years, but one can tell when a nut year is coming up. The year before always sends out a few. The text here is speaking of something quite similar. Most fruit trees yield their first full crop in the fourth year. This first fruit, as also a firstborn male child, belongs to god. The child is to be redeemed, but the fruit belongs to the Priests and the Levites: "Everyone who is clean in your house may eat it." (Lev. 27:30, Num. 18:13)

The fruit of the third year tend to be meager and of poor quality. Growers pick them early because they weaken the fruit to come. Similarly, there is something covering and hiding our heart that prevents it from doing its best. In this way, the unwanted fruit of the third year becomes a corporeal metaphor. We as humans have always wondered: Is there a darkness deep within our souls or is there a *satan* outside of us, boring his way into our souls? Which is the more fruitful metaphor? Our text is looking for a middle way. In each case, foreskin or heart or tree, there is a part that is not part of its integrity. Something on the surface obscures its wholeness—what the text often calls its *completeness.*

> When Abram was ninety-nine years old the Lord appeared to Abram and said to him, "I am God Almighty; walk before me, and be complete. [Gen. 17:1]

The kind of completeness or integrity God asked for was the wholeness achieved by removing and revealing. God proposes to raise up a people through Abram. His terms turn the metaphor back into an action, and then the action is turned back into a sign:

> Every male among you shall be circumcised. You shall be circumcised (וּנְמַלְתֶּם) in the flesh of your foreskins (עָרְלַתְכֶם), and it shall be a sign of the covenant between me and you. [Gen. 17:11]

The Hebrew words מוֹל and עָרְלָה, unlike their English counterparts *circumcise* (around-cut) and *foreskin,* are themselves roots and not derived from simpler elements. That fact gives them a certain amount of power in the language.

The relationship between symbol and symbolized is often more interesting than confusing. I have always thought it interesting that, whereas the Greek word λόγος (*logos*) blurs the distinction between word and thought, the Hebrew דָּבָר (*davar*) blurs the distinction between word and thing. We do the same in the phrase, "Say something." Can we say *things*?

God ended his proposal with a warning:

> Any uncircumcised male who is not circumcised in the flesh of his foreskin shall be cut off from his people; he has broken my covenant. [Gen. 17:14]

And even the stranger is to be circumcised if he is to keep the Passover (Exod. 12:48).

The topic of circumcision will reappear in Exodus, but only after these verses:

> Then you shall say to Pharaoh, "Thus says the Lord, Israel is my firstborn son, and I say to you, Let my son go that he may serve me. If you refuse to let him go, behold, I will kill your firstborn son." [Exod. 4:22–23]

Moses has now been told for the first time that his mission will involve many deaths. Some of those who will die are old men—among them perhaps even some of those who advised Pharaoh to kill the Hebrew children. But others will be newborns. Such deaths might count only as collateral damage for another nation, but for Israel they would establish a debt to be paid. The debt was not one Israel acquired for herself, it was incurred by God on Israel's behalf; nevertheless, it had to be paid. Initially the payment fell upon all firstborns, then only upon the Levites; ultimately it will all fall upon Aaron:

> And the Lord spoke to Moses, saying, "Behold, I have taken the Levites from among the people of Israel instead of every firstborn who opens the womb among the people of Israel. The Levites shall be mine, for all the firstborn are mine. On the day that I struck down all the firstborn in the land of Egypt, I consecrated for my own all the firstborn in Israel, both of man and of beast. They shall be mine: I am the Lord." [Num. 3:11–13]

The verses in Exodus about the killing of all the Egyptian firstborn are immediately followed by the incident at the inn concerning the circumcision of one of Moses' sons; we are not told which son it was:

> At a lodging place on the way the Lord met him [Moses] and sought to put him to death. Then Zipporah took a flint and cut off her son's foreskin and touched his feet with it and said, "Surely you are a bridegroom of blood to me!" So he let him alone. It was then that she said, "A bridegroom of blood," because of the circumcision. [Exod. 4:24–26]

Evidently Moses, on his way to Egypt, stopped off at an inn where he could have, and perhaps even should have, circumcised his son, but for whatever reason did not do so. It is also clear that God judged Moses' behavior to have been worthy of death no matter what his reason may have been.

Some say that Moses put off the circumcision because he was traveling; but that argument never seemed convincing to me, since we read in Joshua that although all the people who came out of Egypt had been circumcised, "the people who were born in the wilderness on the way as they came forth out of Egypt were not circumcised" (Josh. 5:5); and God saw nothing improper in that. Others say that either Moses or the baby was sick. Still others hold that Moses put off the circumcision because he busied himself first with the inn—attending to the baggage and such. But all tacitly agree that, at least up to this point, the text has not told us why Moses did not circumcise his son and therefore we do not know what it was that God thought to be worthy of death.

We have already considered the verse: "Any uncircumcised male who is not circumcised in the flesh of his foreskin shall be cut off from his people; he has broken my covenant." While that does seem to go in the right direction, it does not provide a great deal of help. But I believe we can obtain more guidance from Chapter Twelve of Numbers, in which Miriam and Aaron revolt over Moses' second marriage:

> And Miriam and Aaron spoke against Moses because of the Cushite woman whom he had married; for he had married a Cushite woman. [Num. 12:1]

Before trying to understand Moses' second marriage it would be best to consider his first one and how it compares to similar circumstances surrounding the marriage of Jacob. The following four passages, taken together, invite us to compare the marriages of both characters; I will comment upon them together as a whole.

> And [Rebekah] sent and called Jacob her younger son and said to him, "Behold, your brother Esau comforts himself by planning to slay you (לְהָרְגֶךָ). Now therefore, my son, obey my voice; and arise, flee (בְּרַח) to Laban my brother, to Haran. [Gen. 27:42–43]

> And Jacob went out from Beer-sheba, and went toward Haran. And he came upon the place, and tarried there all night because the sun was set; and he took one of the stones of the place and put it under his head, and lay down in that place to sleep. And he dreamed, and behold, a ladder set up on the earth, and the top of it reached to heaven; and behold, the angels of God ascending and descending on it. And behold, the Lord stood beside him, and said, "I am the Lord, the God of Abraham thy father, and the God of Isaac. The land whereon you lie, to you will I give

> it, and to your seed. And your seed shall be as the dust of the earth, and you shall spread abroad to the west, and to the east, and to the north, and to the south; and in you and in your seed shall all the families of the earth be blessed." [Gen. 28:10–14]
>
> Then Jacob lifted his feet and came to the land of the children of the east. And he looked, and behold, a well (בְּאֵר) in the field; and lo, three flocks of sheep lying there by it, for out of that well they watered the flocks. And the stone upon the well's mouth was great. And thither were all the flocks gathered; and they rolled the stone from the well's mouth, and watered the sheep, and put the stone back upon the well's mouth in its place. And Jacob said to them, "My brothers, where are you from?" And they said, "We are from Haran." And he said to them, "Do you know Laban the son of Nahor?" And they said, "We know him." And he said to them, "Is it well with him?" And they said, "It is well; and behold, Rachel his daughter comes with the sheep." And he said, "Lo, it is yet high day, neither is it time that the cattle should be gathered together; water the sheep, go and feed them." And they said, "We cannot, until all the flocks be gathered together, and they roll the stone from the well's mouth; then we water the sheep." While he was yet speaking with them, Rachel came with her father's sheep; for she tended them. And it came to pass, when Jacob saw Rachel the daughter of Laban his mother's brother, and the sheep of Laban his mother's brother, that Jacob went near and rolled the stone from the well's mouth, and watered the flock of Laban his mother's brother. And Jacob kissed Rachel, and lifted up his voice and wept. And Jacob told Rachel that he was her father's brother, and that he was Rebekah's son; and she ran and told her father. [Gen. 29:1–12]
>
> Now when Pharaoh heard this thing, he sought to slay (לַהֲרֹג) Moses. But Moses fled (וַיִּבְרַח) from the face of Pharaoh, and dwelt in the land of Midian; and he sat down by a well (הַבְּאֵר). And the priest of Midian had seven daughters; and they came and drew water, and filled the troughs to water their father's flock. And the shepherds came and drove them away; but Moses rose up and helped them, and watered their flock. And when they came to Reuel their father, he said, "How is it that you are come so soon today?" And they said, "An Egyptian delivered us out of the hand of the shepherds, and moreover he drew water for us, and watered the flock." And he said to his daughters, "And where is the man? Why is it that you have left him? Call him, that he may eat bread." And Moses was content to dwell with the man; and he gave Moses Zipporah his daughter. [Exod. 2:15–21]

Jacob and Moses started off with much in common. Both were fleeing (בְּרַח) one who would slay (הָרֹג) them, and both found themselves at the well (בְּאֵר) of a kinsman.[2] Both came upon a group of shepherds, and in each

2 The Midianites were descendants of Abraham (Gen. 25:2).

case the visit led to a marriage. But their stories also display significant contrasts. When Moses arrived at his well, the story was both strangely familiar and strangely unfamiliar. If we compare the accounts of these two men bit by bit, I believe there is much we can learn about Moses.

• Jacob bedded down for the night on his way to Haran and learned in a dream that God would be with him and that his progeny would become a great nation.

But Moses came to Midian straightway in his flight from Pharaoh. For him there was no intervening encounter.

• Jacob became more light-hearted after his dream. His future looked brighter and he seemed to have forgotten all his fears. As the text puts it, "Jacob picked up his feet[3] and came to the land." One can hear this same spirit when he addresses the shepherds as "my brothers" (Gen. 37:4), a term rarely found in Tanar[4] except in the most literal sense; and it would certainly never be used when addressing a stranger. Jacob then asked the shepherds, "Where do you come from?" Through this and other inquiries, he learned that they were from Haran and that they knew his uncle Laban well enough to say that "it is well with him"; and we learn that the shepherds are close enough to Laban to know how he is faring.

But Moses made no such inquiries even though, as we shall learn later, there were some shepherds about.

• Jacob heard from the shepherds that Rachel, Laban's daughter, was coming with the sheep. He now knew her name.

To understand the importance of this knowledge, consider the force of names as it reveals itself in Chapter One of Genesis.

Named	*Name*
light	Day
darkness	Night
the spread-out	Sky
the dry	Land
the gathering of the waters	Sea

From this point of view, names give form and limit to the unlimited. Light has no end, but Day begins when Night is over. The spread-out goes on and on, but the Sky is defined by the horizon. Sea and Land come together and are

3 We would say, "picked up his heels."

4 The Hebrew acronym *Tanar* refers to the books of Genesis through II Kings, plus Ruth.

held apart by the shoreline. Names make an it or a what or a who out of an endless mass. To know a name is already to have acknowledged personhood—that which distinguishes one from all others, even if nothing else is known.

But for Moses it was seven nameless shepherdesses who came down to water and filled the troughs to water their father's flock. For him there were no names, just a mass of women.

- Jacob did his best to set the stage by trying to persuade the shepherds to water their sheep and go, leaving the scene for a first meeting. When that proved not to be feasible he devised an alternative plan. He rolled away the great stone that was to be rolled away only after all the shepherds had gathered. To me it is not clear whether it required great physical strength to roll the stone away or whether there was a common agreement or law that no one would open the well until all were present so that no one might have more than his fair share of the water. Either way, as strong man or rule-breaker, Jacob made a great show of gallantry.

  But Moses found a different world around the well in Midian. While the shepherds of Haran were peaceful even when Jacob acted against their custom and perhaps their law, the shepherds of Midian drove away the daughters of Reuel until Moses rose up to save them.

- Both Jacob and Moses watered a flock of sheep.

  But Moses faced a crowd of angry men to save seven shepherdesses whom in all probability he would never see again.

- Jacob revealed himself to Rachel as her father's brother.

  But Moses said nothing of his family.

- Rachel ran to tell Laban her father.

  But the sisters went home and said nothing of their meeting with the stranger until their father, Reuel, made inquiry. When his daughter told him what had happened, Reuel said: "Then where is he? Why have you left the man? Call him, that he may eat bread" (Exod. 2:20).

- Laban married off his elder daughter Leah in place of Rachel.

  Reuel married off one of his daughters to Moses, and he will bring her and Moses back together when Moses is in the wilderness (Exod. 18:5); but in neither case is the union long lasting.

- Jacob lived most of his life with two wives.

  But Moses lived most of his life with none, so far as we know. Aside from the incident at the inn we never see any interaction between Moses and either of his wives.

• Jacob will work fourteen years to gain Rachel as his wife, "and they seemed to him but a few days, for the love he had for her" (Gen. 29:20).

But Moses was simply given a wife, and we are told nothing of how she was selected or what part he may or may not have played in the arrangement.

The question of Moses' relationship to his wives will be taken up later in this discussion. But let us recall that the passage in Numbers which prompted us to examine Moses' earlier marriage begins: "Miriam and Aaron spoke against Moses because of the Cushite woman whom he had married; for he had married a Cushite woman." Superficially, Miriam and Aaron have a point: Moses already had a wife, Zipporah, a direct descendent of Abraham through Isaac's younger half-brother, Midian (Gen. 25:1–2). Not only was this new Cushite wife a descendent of Ham, the problem was even more involved.

Cush fathered Nimrod, the first on earth to be a hero. He was the source of the Heroic Cult, the main rival to the teaching of Moses. This cult would include not only Gilgamesh but even Yavan[5] and beyond. All of Israel's future captors—Babylon, Assyria, Nineveh—would become Nimrod's kingdom (Gen. 10:6–14). Why would Moses marry this woman?

God did not directly address the charges which Miriam and Aaron laid against Moses, saying instead:

> "Hear my words: If there is a prophet among you, I the Lord make myself known to him in a vision; I speak with him in a dream. Not so with my servant Moses. He is faithful in all my house. With him I speak mouth to mouth, clearly, and not in riddles, and he beholds the form of the Lord. Why then were you not afraid to speak against my servant Moses?" [Num. 12:6–8]

Aaron and Miriam were right in holding that God does not speak through Moses alone. There were and would continue to be many prophets; but Moses was not a prophet, strictly speaking. God reveals thoughts to the prophets through the imagination. The prophets grasp the world through similes, or they work out their glimpses in a dream; but the phrase "mouth to mouth" implies that God speaks directly to Moses' intellect rather than through his imagination. The magnitude and growth of the distinction between Moses and the prophets can best be seen by comparing the following two verses:

> And he said, "I am the God of your father, the God of Abraham, the God of Isaac, and the God of Jacob." And Moses hid his face, for he was afraid to look at God. [Exod. 3:6]

---

5 Yavan (יָוָן): Ionia; later, Greece as a whole.

and

> When Moses came down from Mount Sinai, with the two tablets of the testimony in his hand as he came down from the mountain, Moses did not know that the skin of his face shone because he had been talking with God. And when Aaron and all the people saw Moses, behold, the skin of his face shone; and they were afraid to come closer to him.... And when Moses had finished speaking with them, he put a veil on his face. Whenever Moses went in before the Lord to speak with him, he would remove the veil, until he came out. And when he came out and told the people of Israel what he was commanded, the people of Israel would see the face of Moses, that the skin of Moses' face was shining. And Moses would put the veil over his face again, until he went in to speak with him. [Exod. 34:29–34]

The people could not look for long on Moses' face, even as Moses was afraid to look upon God. On what a knife-edge the whole stands. There Moses emerged, set apart from all other men. How like a god he had to become if his teaching was to endure; yet if he had actually become a god, paganism with its need for visible divinity would have returned, and all would have been lost.

PART THREE

Israel comprised twelve tribes with but one altar. From the Tent of Moses to the Temple of Solomon, that single altar was alone the place of the appointed joyous occasions that held the nation together. Not a king, but a sacred structure was to unify the twelve.

> And you shall rejoice in your feast, you and your son and your daughter, your male servant and your female ... because the Lord your God will bless you in all your produce and in all the work of your hands, so that you will be altogether joyful. Three times a year all your males shall appear before the Lord your God.... Every man shall give as he is able, according to the blessing of the Lord your God that he has given you. [Deut. 16:14–17]

The tribes far to the North knew it; and they understood that while it might be duplicated as a memorial, no other altar was to be used for worship.

> ...that your children will not say to our children in time to come, "You have no portion in the Lord." And we thought, If this should be said to us or to our descendants in time to come, we should say, "Behold, the copy of the altar of the Lord, which our fathers made, not for burnt offerings, nor for sacrifice, but to be a witness between us and you." [Josh. 22:27–28]

Even those who revolted against it knew it.

> And Jeroboam said in his heart, "Now the kingdom will turn back to the house of David. If this people go up to offer sacrifices in the house of the Lord at Jerusalem, then the heart of this people will turn again to their Lord, to Rehoboam king of Judah, and they will kill me and return to Rehoboam king of Judah." So the king took counsel and made two calves of gold. And he said to the people, "You have gone up to Jerusalem long enough. Behold your gods, O Israel, who brought you up out of the land of Egypt." [I Kings 12:26–28]

The Lord had intended the Temple in Jerusalem to be the sole means of unifying the tribes of Israel. But that plan would fail, as the book of Judges recounts. The story begins this way:

> There was a man of the hill country of Ephraim, whose name was Micah. He said to his mother, "The eleven hundred pieces of silver that were taken from you and is with me, about which you uttered a curse, and spoke it in my ears, behold, the silver is with me; I took it." And his mother said, "Blessed be my son of the Lord." [Judg. 17:1–2]

Micah began by speaking in the passive voice. There is always something suspicious about that voice, since it avoids the notion of cause and hence also any notion of responsibility. Note the chiastic form of his speech:

Eleven hundred pieces of silver	He states the conditions
were taken from you	No responsibility
and is with me	Admits some relationship
you uttered a curse	Curse (general)
spoke it in my ears	Implied accusation (specific)
the silver is with me	Admits some relationship
I took it.	Full responsibility
Blessed be my son of the Lord.	She states the results

The process by which Micah's mother transformed Micah's passive voice into his confession is worthy of note. However it is not clear whether her last words are a blessing or a euphemistic curse. The text continues:

> He restored the eleven hundred pieces of silver to his mother. And his mother said, "I dedicate the silver to the Lord from my hand for my son, to make a carved image (פֶּסֶל) and a molten image (וּמַסֵּכָה). Now therefore I will restore it to you." So when he restored the money to his mother, his mother took two hundred pieces of silver and gave it to the silversmith. [Judg. 17:3–4a]

Notice the difference between the amount pledged and the amount spent. Also note the discord between the dedicated objects and the god to whom they are dedicated:

> [The] silversmith made it into a carved image (פֶּסֶל) and a molten image (וּמַסֵּכָה). And it was in the house of Micah. The man Micah had a house of God (בֵּית אֱלֹהִים), and he made an ephod (אֵפוֹד) and household gods (וּתְרָפִים), and ordained (וַיְמַלֵּא אֶת־יַד) one of his sons, who became his priest (כֹּהֵן). [Judg. 17:4b–5]

Micah has jumbled together objects like *beth-el*, *ephod*, and *priest*, associated with the God of Israel, and articles like *carved image*, *molten image*, and *household gods* that belong to the pagan world. For recall the following passages:

> You shall not make for yourself a *carved image* (פֶּסֶל), or any likeness of that is in heaven above, or that is in the earth beneath, or that is in the water under the earth. [Exod. 20:4]

> And he received [the gold] from their hand and fashioned it with a graving tool and made a *molten* (מַסֵּכָה) calf. And they said, "These are your gods, O Israel, who brought you up out of the land of Egypt." [Exod. 32:4]

> "You shall not make for yourself any *molten* (מַסֵּכָה) gods." [Exod. 34:17]

> Then Jacob awoke from his sleep and said, "Surely the Lord is in this place, and I did not know it." And he was afraid and said, "How awesome is this place! This is none other than the house of God, (בֵּית אֱלֹהִים, *beth elohim*) and this is the gate of heaven." [Gen. 28:16–17]

> And they shall make the *ephod* (אֵפֹד) of gold, of blue and purple and scarlet yarns, and of fine twined linen, skillfully worked. [Exod. 28:6]

> Laban had gone to shear his sheep, and Rachel stole her father's *household gods* (תְרָפִים). [Gen. 31:19]

> And you shall put them on Aaron your brother, and on his sons with him, and shall anoint them and ordain (וּמִלֵּאתָ אֶת־יָדָם) them and consecrate them, that they may be *priests* (וְכִהֲנוּ) to me. [Exod. 28:41]

It is now evident that this strange amalgam of sacred objects—some pagan, others from the tabernacle itself—was a temple of confusion. Micah's shrine was an attempt to drag the God of Israel into the pagan world, even as Aaron's altar—built in front of the idol—and his call for a feast to the Lord had been his attempt to draw the people back from that world (Exod. 32:3–5).[6]

The rest of the story of Micah begins with a haunting phrase that will appear and reappear and finally bring the book of Judges to its fated conclusion: "In those days there was no king in Israel. Everyone did what was right in his own eyes" (Judg. 17:6, 18:1, 19:1, 21:25). The story continues, telling of a young man who came to the house of Micah:

---

6 As we shall see, a successor to Micah's shrine will call forth the very words that were used on that earlier occasion: "These are your gods, O Israel, that brought you up out of the land of Egypt."

> Now there was a young man of Bethlehem in Judah, of the family of Judah, who was a Levite; and he stayed (גָר) there. And the man departed from the town of Bethlehem in Judah to stay wherever he could find a place. He came to the hill country of Ephraim, to the house of Micah, as he was making his way. Micah said to him, "Where do you come from?" And he said to him, "I am a Levite from Bethlehem in Judah, and I will stay wherever I find a place." [Judg. 17:7–9]

This nameless young man appeared to be something of a rootless fellow. He was a Levite but had some connection to the family of Judah. He "stayed" in Bethlehem but does not seem to have lived there. It is difficult to understand the words "of the family of Judah." The best I can make of them is to assume that although he was a Levite, he spent most of his time in the surrounding Judean community and did not feel himself at home in the small Levite community. The wanderings of this nameless man seem to have brought him to the house of Micah by sheer chance.

> And Micah said to him, "Settle down (שְׁבָה) with me, and be a father and a priest to me, and I will give you ten pieces of silver a year and a suit of clothes and your living." And the Levite went in. [Judg. 17:10]

Notice how easily the Levite accepts the change from "stay" (גָר) to "settle down," (שְׁבָה) perhaps without even noticing the difference. We will see how seriously he takes his own word and how well he keeps it.

> And the Levite was content to settle down with the man, and the young man became to him like one of his sons. And Micah ordained the Levite, and the young man became his priest, and was in the house of Micah. Then Micah said, "Now I know that the Lord will prosper me, because I have a Levite as priest." [Judg. 17:11–13]

One is struck by the conflict one must feel between Micah's pride and the reader's horror at seeing an ordinary Levite become a priest. The text goes on to summarize the situation by saying, for the second time, "In those days there was no king in Israel. Everyone did what was right in his own eyes" (Judg. 17:5–6). To understand the force of these words, we must recall some earlier incidents.

Shortly before his death, Moses said to the Lord,

> "Let the Lord ... appoint a man ... who shall lead them out and bring them in, that the congregation of the Lord may not be as sheep that have no shepherd. The Lord answered, "Take Joshua the son of Nun." [Num. 27:16–18]

Joshua himself never made such a request; but after his death

> The people of Israel inquired of the Lord, "Who shall go up first for us against the Canaanite?" The Lord said, "Judah shall go up." [Judg. 1:1–2]

The Hebrew word translated "shall go up" is a third person singular verb. But Judah the man had been dead for over three hundred years; God was speaking of the whole tribe as if they were one single individual. In the same vein,

> Judah said to Simeon his brother, "Come up with me." [Judg. 1:3]

Again all is in the singular. There would be no king; all would go as one. The book of Proverbs gives a clear picture of what that idyllic world would have been like:

> The locusts have no king, yet all of them march in rank. [Prov. 30:27; see also Prov. 6:6–11]

The book of Judges begins in that idyllic world—the same world Gideon had in mind when he said: "I will not rule over you, and my son will not rule over you; the Lord will rule over you" (Judg. 8:23), or what Samuel meant when he said:

> When you saw that Nahash the king of the Ammonites came against you, you said to me, "No, but a king shall reign over us," when the Lord your God was your king. [I Sam. 12:12].

The book of Judges is an account of that idyllic world, and of its inevitable devolution into total chaos.

In the chapter that follows, Micah will face men from the tribe of Dan. The first verse opens with a familiar theme:

> In those days there was no king in Israel. And in those days the tribe of the people of Dan was seeking for itself an inheritance to dwell in, for until then no inheritance among the tribes of Israel had fallen to them. [Judg. 18:1]

There were two main reasons for this failure. As the reader will recall from the first part of our discussion, Israel was to have entered the promised land from the south. Had that happened, Dan would have been the first to enter the land, and all the troops of Israel would have been with them. But as it actually happened, the tribe of Dan was last to enter; all the other Israelites were off establishing their own territories and there was no one to serve as ally to Dan.

Tribe	Sons	Men of War	Men/Sons
Reuben	4	46.500	11,625
Simeon	6	59,300	9,883
Levi	3		
Judah	5	74,600	14,920
Issachar	4	54,400	13,600
Zebulun	3	57,400	19,333

Gad	7	45,650	6,521
Asher	4	41,500	10,375
Joseph	2		
Ephraim	3	40,500	13,500
Manasseh	4	32,200	8,050
Benjamin	10	35,400	3,540
Dan	1	62,700	62,700
Naphtali	4	53,400	3,337

As the reader can see from this chart of the sons of Jacob, Dan alone had only one son. Yet when the time came to settle the territory, his tribe had the greatest population except for Judah. The land was too small; they had to move. For this and other reasons, they began to look to the north for a new land.

> The sons of Dan sent out five men to scout out the land and explore it.... When they were by the house of Micah, they recognized the voice of the young man, the Levite; and they asked him, "Who brought you here? What are you doing in this place? And what is here for you?" [Judg. 18:2–3]

With the words "The sons of Dan," the text presents the Danites as having no king. They were still following that idyllic way of life even after the first verse's foreshadowing that such a life—without a king—would soon end in gruesome chaos. How are we to understand the fact that the scouts from Dan recognized the voice of this so far lackluster young Levite? It would also be interesting to compare this scouting expedition with the one sent out by Moses so many years before.

> [The Levite] said to them, "Micah did this-and-that to me, and he has hired me; and I have become his priest." [Judg. 18:4]

It is not clear what ceremony was undertaken when the Levite became a priest, but he makes it sound fairly banal compared to the priestly ordination described in the Torah (Exod. 29:9).

The Danites quickly accept the Levite as a priest and ask him to inquire of God, "that we may know whether the journey on which we are setting out will succeed" (Judg. 18:4); but it is not clear whether they truly want a priest or whether they just want to know where they stand in the eyes of the Levite.

> The priest said to them, "Go in peace. The journey on which you go is in accord with the Lord." [Judg. 18:6]

Readers encountering this story for a second time and who therefore know what is to come are horrified by this passage. Its strength lies in the tension that arises between the words of the priest and the feelings they bring out in the reader who knows that the sequel will be far from peaceful and hardly pleasing to the Lord.

There are two separate accounts of the next part of our story. One is a rather straightforward narration to be found in the book of Joshua:

> The seventh lot came out for the tribe of the people of Dan, according to their clans. And the territory of its inheritance included Zorah, Eshtaol, Ir-shemesh, ... Timnah, Ekron.... When the territory of the people of Dan was lost to them, the people of Dan went up and fought against Leshem (לֶשֶׁם); and after capturing it and striking it with the sword they took possession of it and settled in it, calling Leshem Dan, after the name of Dan their ancestor. [Josh. 19:40–47]

The second account is a retelling of the first:

> Then the five men departed and came to Laish and saw the people who were there, how they lived in security after the manner of the Sidonians, quiet and unsuspecting, no one brought shame upon the land by taking possession of (יוֹרֵשׁ) the power to restrain. They were far from the Sidonians and had no dealings with anyone. [Judg. 18:7]

The people of Laish quietly live the idyllic life that Israel might have enjoyed had life without a king been possible for them. The people of Laish are well described as the *ant* in the verses: "Go to the ant, O sluggard (עָצֵל); consider her ways, and be wise. Without having any chief, officer, or ruler, she prepares her bread in summer and gathers her food in harvest" (Prov 6:6–8). The main teaching of the book of Judges, however, is the fragility of the idyllic. The account continues:

> When [the scouts] came to their brothers at Zorah and Eshtaol, their brothers said to them, "What do you have?" They said, "Get up, let us go up against them. We have seen the land, and it is very good. You just stand there. Do not be sluggards (אַל־תֵּעָצְלוּ); go in and take possession of (לָרֶשֶׁת) the land." [Judg. 18:8–9]

Note that in this chapter two peoples will meet. Each is a people that has no king. Among them, "everyone did what was right in his own eyes." But one was like the ant, the other was not. The scouts have urged their brothers to commit the very act that the people of Laish said that they would never commit: taking possession (ירש) of another's land. By the end of the book both people will be decimated, one from outside, the other from inside.

> So six hundred men of the tribe of Dan, armed with weapons of war, set out from Zorah and Eshtaol, and went up and encamped at Kiriath-jearim in Judah.... They passed on ... and came to the house of Micah. Then the five men who had gone to scout out the country of Laish said to their brothers, "Do you know that in these houses are an ephod,

> household gods, a carved image, and a molten image? Now therefore consider what you will do." And they turned aside there and came to the house of the young Levite, at the home of Micah, and asked him about his welfare. Now the six hundred men of the Danites, armed with their weapons of war, stood by the entrance of the gate. [Judg. 18:11–16]

The brothers answered the call with "six hundred men of the tribe of Dan, armed with weapons of war," from Zorah and Eshtaol (Judg. 18:11). How different were these men from Caleb's fainthearted fellow-scouts (Num. 13:30–31)!

The Danites' boldness may stem in part from their having come directly from the funeral of Samson, as mentioned in the verse immediately preceding the story of Micah:

> Then [Samson's] brothers and all his family came down and took him and brought him up and buried him between Zorah and Eshtaol in the tomb of Manoah his father. [Judg. 16:31]

The five men are never referred as "commanders" of the Danites, but simply as those who had gone to scout out the country. The six hundred are not "their army" but "their brothers." The five give their brothers the information they need to make a decision, and they ask them to consider what to do. They indeed have no king.

> And the five men who had gone to scout out the land went up and entered and took the carved image, the ephod, the household gods, and the molten image, while the priest stood by the entrance of the gate with the six hundred men armed with weapons of war. And when these went into Micah's house and took the carved image, the ephod, the household gods, and the molten image, the priest said to them, "What are you doing?" [Judg. 18:17–18]

The interplay between the main action and the six hundred standing by the gate is chilling.

> And they said to him, "Keep quiet; put your hand on your mouth and come with us and be to us a father and a priest. Is it better for you to be priest to the house of one man, or to be priest to a tribe and clan in Israel?" And the priest's heart was glad. He took the ephod and the household gods and the carved image and went along with the people. [Judg. 18:19–20]

At the thought of becoming a priest to a whole tribe, our passionless Levite has suddenly come alive; his "heart was glad" (וַיִּיטַב לֵב). The expression is rather rare in Tanakh. Absalom told his servants to *strike Amnon when his heart is glad with wine* (II Sam. 13:28). But it is also what the people felt on the day that Solomon dedicated his Temple (I Kings 8:66).

The Danites seized the shrine objects and marched off, putting the little ones, the livestock, and the goods in front of them. Normally, children and wives would be in the rear for protection; but the Danites know that Micah and his neighbors would be faster and catch them from behind.

> When they came, the people of Dan turned around and said to Micah, "What is the matter with you, that you come with such a company?" And he said, "You take my gods that I made and the priest, and go away, and what have I left? How then do you ask me, 'What is the matter with you?" [Judg. 18:21–24]

It is difficult not to feel sympathy for the victim of such a brazen crime; his words, "that I made," give even more force to his plaint. It seems that even in a case like this, we so need representations of *good vs. bad*, it takes a moment or two to remember that this is a case of *bad vs bad.*

> The people of Dan said to him, "Do not let your voice be heard among us, lest angry fellows fall upon you, and you lose your life with the lives of your household." [Judg. 18:25]

Few things are more intimidating than a friendly warning that a third party is out to do one harm, when both know that there is in fact no third party.

> Then the people of Dan went their way. And when Micah saw that they were too strong for him, he turned and went back to his home. And the people of Dan took what Micah had made, and the priest who belonged to him, and they came to Laish, to a people quiet and unsuspecting, and struck them (וַיַּכּוּ) with the edge of the sword (לְפִי חָרֶב) and burned (שָׂרְפוּ) the city (הָעִיר) with fire (בָאֵשׁ). And there was no deliverer because it was far from Sidon, and they had no dealings with anyone. It was in the valley that belongs to Beth-rehob. Then they rebuilt the city and lived in it. [Judg. 18:26–28]

In some misguided way the Danites have seen their kinship with the people of Laish, that other people with no king, and applied to them a contorted form of the enjoinder

> If you hear in one of your cities ... that certain worthless fellows have gone out ... saying, "Let us go and serve other gods," ... you shall surely strike (הַכֵּה תַכֶּה) the inhabitants of that city with the edge of the sword (לְפִי חָרֶב).... You shall burn (וְשָׂרַפְתָּ) the city (הָעִיר) and all its spoil with fire (בָאֵשׁ).... It shall not be built again. [Deut. 13:13–17]

But if so, Dan has rebuilt that which, according to their own interpretation of this passage, should not be built again.

> And they named the city Dan, after the name of Dan their ancestor, who was born to Israel; but the name of the city was Laish (לַיִשׁ) at first. [Judg 18:29]

When Moses said, "But now, if you will forgive their sin—but if not, please blot me out of your book that you have written" (Exod. 32:32), or when God said, "Let me alone, that I may destroy them and blot out their name from under heaven. And I will make of you a nation mightier and greater than they" (Deut. 9:14), both speakers envision the obliteration of any sign that the person or nation ever existed. Indeed, how many place names there are in the world that go back to dead languages where nothing else is remembered! But if annihilation of the idyll is what Dan intended, the narrator seems to have prevented that by tacking on the concluding phrase, "but the name of the city was Laish (לַיִשׁ) at first." Or was it Leshem (לֶשֶׁם)?

PART FOUR

Before we turn to the last two verses of our chapter, we should review a few particulars concerning the life of Moses:

- The occasion of his first marriage was quite different from Jacob's. Had it not been for Reuel, Moses might never have seen any of the seven sisters, Zipporah included, again. We have no idea why Zipporah was chosen to be his wife, or whether Moses was consulted when the choice was made.

- In the rather complicated account rehearsing the names of the sons of Levi, their wives and sons as well as the names of their sons' wives and sons—some thirty-odd names in all—neither the names of Moses' wife nor the names of his sons ever appear.

- Moses and his wife were together at the inn, but things did not go well. Since Moses did not circumcise the boy, she did—although her feelings about the matter are not quite clear. As we learn later, Moses, while still at the inn, sent her away (שִׁלּוּחֶיהָ), back to Media, where she remained with the children until her father brought her to Moses at the camp in the wilderness (Exod. 18:2–5), just as he had done on the day they first met. But after her father (now called Jethro) left the camp she was never heard of again. We also learn that Moses took a second wife, one from the land of Cush, progenitor of all of Israel's captors; but we do not know why he did so.

- Moses' burial place is unknown:

> He buried him in the valley in the land of Moab opposite Beth-peor; but no one knows the place of his burial to this day. [Deut. 34:6]

- Most importantly,

> The man Moses was very meek (עָנָיו) more than all people who were on the face of the earth. [Num. 12:3]

The word עָנָיו, here translated *meek,* appears often in the poetical books and in the books of the Later Prophets, where it consistently means either *poor* or *afflicted.* In our verse, however, the word has a particularly striking force, since this is the only time it occurs in the *Tanar,* that is, the books of Genesis through II Kings. The unique appearance of this word reveals the great contrast between the humility of the man and the grandeur of the office he is to fill.

The last verse but one of our chapter in Judges may shed some light on what it was that happened at the inn:

> The people of Dan set up the carved image (פֶּסֶל) for themselves; and Jonathan, son of Gershom son of Moses, and his sons, were priests to the tribe of the Danites until the day of the exile of the land (עַד־יוֹם גְּלוֹת הָאָרֶץ). [Judg. 18:30]

What Moses feared most has come to pass. He foresaw the temptation for one of his descendants to regard himself as the true heir to Moses and the priesthood; exactly as Jonathan has done. Having no reason to believe that his descendants would inherit his meekness, Moses determined to leave no heir:

- He did not accompany the sisters when they returned to Jethro.
- He did not seek a wife but was given one.
- Moses and Zipporah quarreled over the circumcision of their son.
- After the affair at the inn, wife and sons were sent back to Midian; and when Jethro later brought them back to Moses they were totally ignored.
- When Moses undertook a second marriage, he chose a woman whose progeny would be least able to assume leadership.

Nor would Moses leave a symbolic presence:

- The names of those who would have been his heirs do not appear in the detailed account of the sons of Levi.
- Moses' burial place is unknown.

All this leads to the suggestion that Moses never intended to circumcise his son; that way Moses could disappear one day without an heir, ending the Mosaic line. No one would claim to have inherited the position of lawgiver, and no one would be worshiped as a god. But why, then, did God threaten to kill Moses? Why did God oppose such a well-thought-out plan? That question must be left to the reader to ponder.

The reader who has access to the previous passage through translations alone may well be confused, since some translators render *Jonathan, son of Gershom son of Moses,* while others have *Jonathan, son of Gershom son of Manasseh.* Why the discrepancy? For those who do not read Hebrew,

Moses is written	משה
Manasseh is written	מנשה
But the biblical text reads	מ^נשה

The text incorporates a superscript נ (*nun*) which, if ignored, yields *Moses*, but if treated as normal letter yields *Manasseh*. Since Moses did have a son named Gershom, whereas Manasseh did not, it seems certain that the original text read *Moses*. But perhaps a later generation of scribes, thinking it unseemly that Moses should have a grandson like Jonathan, changed the name to *Manasseh*. A third generation of scribes then came along. Either they were not sure which text was correct but thought *Moses* the more likely, or they discovered the change and restored the text, but for one reason or another wished to retain a sign that such a change had once been made.

The last two verses of our inquiry read:

> And the people of Dan set up the carved image for themselves, and Jonathan, the son of Gershom son of Moses, and his sons were priests to the tribe of Danites until the day of the exile of the land. So they set up Micah's carved image that he made, as long as the house of God was at Shiloh. [Judg. 18:31]

Thus the shrine at whose setting-up Jonathan officiated as priest was essentially Micah's, since it held all the sacred utensils that were stolen from Micah's shrine. But what shall we make of the phrase: "until the day of the exile of the land" (עַד־יוֹם גְּלוֹת הָאָרֶץ)"? Clearly it cannot mean that rocks and rivers and mountaintops can be carried off to a foreign land. However *the land* (הָאָרֶץ) can also refer to *the people of the land*, as in the verse:

> The land (הָאָרֶץ) shall mourn, each family by itself: the family of the house of David by itself, and their wives by themselves; the family of the house of Nathan by itself, and their wives by themselves. [Zech. 12:12]

When the word הָאָרֶץ is used in this way it tends to stress *completeness*: as we might say, *the country rejoiced*. On the other hand, when the text speaks not of the land that went into exile but the land they were forced to leave behind, it uses the unrelated term אָדָם (*adam*), something one can grab by the handful:

> So Israel was exiled (וַיִּגֶל) from its land (אַדְמָתוֹ) to Assyria until this day. [II Kings 17:22–23]

It seems clear, then, that the shrine in Dan lasted until the time of the Assyrian exile. At any rate, that is the first time any verb having the root גלה (in the sense of *exile*) appears in regard to Israel.

Aside from Judah, Dan alone had the knowledge needed to build a *Beth El*, since we read:

> He [God] has inspired him [Bezaleel] to teach, both him and Oholiab the son of Ahisamach of the tribe of Dan. He has filled them with skill to do every sort of work done by an engraver or by a designer or by an embroiderer in blue and purple and scarlet yarns and fine twined linen, or by a weaver—by any sort of workman or skilled designer. [Exod. 34:34–35]

At the beginning of Part Three of this paper we saw the importance to Jeroboam of establishing an alterative to the House of the Lord (*Beth El*) that stood at Jerusalem. For reasons given at the time, one might have expected Jeroboam to establish only a single place of sacrifice, since its function was to unify the tribes and avert any need the people might feel for returning to the temple in Jerusalem. Instead, "he took counsel and made two calves of gold.... And he set one in Bethel, and one he put in Dan" (I Kings 12:28–29). As we have seen, Jonathan's shrine still stood in Dan at this time; and apparently Jeroboam was in no position to neglect it. But so far as we are told, he never visited the shrine in Dan, whereas the most portentous day of his life was spent at Bethel (I Kings 13:1–10).

What Jeroboam did will vibrate throughout the rest of the text. Twenty times we will read the phrase "the sins of Jeroboam, which he sinned and made Israel to sin." The first of these occasions reads:

> The Lord will strike Israel as a reed is shaken in the water, and root up Israel out of this good land that he gave to their fathers and scatter them beyond the Euphrates, because they have made their Asherim, provoking the Lord to anger. And he will deliver Israel over because of the sins of Jeroboam, which he sinned and made Israel to sin. [I Kings 14:15–16]

According to this verse, the sins of Jeroboam—that is, the lasting presence of those two golden calves—were the ultimate cause of the Assyrian exile.

The account in Kings of Jeroboam's offenses continues: "He set one in Bethel, and one he put in Dan; and this thing became a sin, that the people went before the one—to Dan" (I Kings 29–30). Clearly, the true sin was going to Dan. The question remains: Was it because of their zeal that they went all the way to Dan? Or was it because they worshiped at the shrine of Jonathan?

Before going on, there are some things the reader should know about the city of Shiloh, in the territory which would become Ephraim:

> Then the whole congregation of the people of Israel assembled at Shiloh and set up the tent of meeting there. The land lay subdued before them. [Josh. 18:1]

This is the first mention of the city Shiloh. In other words, we have no notion of why Shiloh was chosen. It was Joshua's base as he fixed the tribal borders.

Israel once launched an uncommanded war against the Philistines from there and lost.

> And when the people came to the camp, the elders of Israel said, "Why has the Lord defeated us today before the Philistines? Let us bring the ark of the covenant of the Lord here from Shiloh, that it may come among us and save us from the power of our enemies." So the people sent to Shiloh and brought from there the ark of the covenant of the Lord of hosts, who is enthroned on the cherubim. [I Sam. 4:3–4]

The second battle too was lost, the ark of the Lord captured, and Pinchas and Hophni, sons of the High Priest, were killed in the battle. The High Priest himself died of shock, and Pinchas's wife lamented:

> "The glory has departed (גָּלָה) from Israel, for the ark of God has been captured." [I Sam. 4:22]

The ark was ultimately returned, but it was never again to reside in Shiloh.

With all this from the first book of Samuel in mind, let us recall the final verse of our chapter from the book of Judges:

> So they set up Micah's carved image that he made, as long as the house of God was at Shiloh. [Judg. 18:31]

According to this verse, Jonathan's shrine was destroyed near the end of the time of the Judges, about three hundred sixty years before the Assyrian exile. Temporally, the two accounts are incompatible. Nor do I see any way of considering either the departure of glory or the capture of the ark as being compatible with "the exile of the land." Perhaps whoever made the change from *Moses* to *Manasseh* also thought to minimize the longevity of Jonathan's shrine by replacing the verse concerning the Assyrian exile with the verse concerning Shiloh—a contradiction subsequently left intact, perhaps by the same hand that later set the superscript *nun*.

Samson's Riddle

> And the Lord said: "I will destroy man whom I have created from off the face of the earth: both man and beast and creeping things, and the birds of the air: for I repent that I have made them." But Noah found favor in the eyes of the Lord. [Gen. 6:7,8]

This much we must always remember: if God had not noticed Noah, the fish, not Israel, would have been God's chosen: they, not man, would have been the summit of creation. These two, fish and man, had been linked in many ways even from the beginning. They were both "created"; all other things were "made" (Gen.1:21, 27). Only they received a blessing (Gen. 1:22, 28); and though the fish, but not man, is said to be "good" (טוֹב, Gen. 1:21), neither man nor fish is said to be "so" (כֵּן). On the first half of the sixth day all will return to normal. The animals will be made, not created. They will be "so," but they will not receive a blessing.

At first, it seems an odd pairing. Fish are at home in the chaotic waters of the sea. They know no laws, and they have no king. But we cannot live in the sea; their home would be our death. How different these two, man and the fish! And yet neither man nor fish is said to be "so." What then is meant by the phrase "and it was so"? From the context, it does not seem to mean anything like "as it had been in speech, so is it now in deed," since the phrase usually occurs prior to the actual coming-to-be of the object itself. The Hebrew word כֵּן, if we can judge anything rightly by its roots, seems to mean something like "having a clear way in which to be,"[1] as in the verse

> And Jacob did so (וַיַּעַשׂ יַעֲקֹב כֵּן). [Gen. 29:28]

What is this kinship we have to the denizens of the sea, this strange lack of an "it was so"? They neither follow the ecliptic as does the sun, nor are they restricted to one surface of motion. No paths have been marked out for them in the sea from the beginning. The created beings share an openness of direction, since man too can wander. Perhaps it is because they are both open and free to wander that they are the only two in need of a blessing. And what of that home, the sea? What are its beginnings? Again, everything is just a bit off—nothing is perfectly clear. Normally we say that it was the second day of creation which was devoted to the establishment of the sky and the waters.

---

1 See my remarks on Genesis 1:11 in *The Lion and the Ass: Reading Genesis after Babylon.* Kafir Yaroq Books (2019)

Indeed, this would seem to be quite right, and exactly according to God's great overall plan for creation:

Day One: Light	Day Four: Sun and Moon
Day Two: Sky and Water	Day Five: Birds and Fish
Day Three: Dry Land including Plants	Day Six: Animals including Man

God's plan was clear, simple, and even elegant. The work of the last three days was devoted to the coming-to-be of the manifestly moving beings which inhabit the several kinds of places prepared for them on the first three corresponding days. But things could not go so smoothly, as the author indicates by not including the words "and the Lord saw that it was good" in the account of the second day. The brute fact is that God's beautiful plan could not work. There cannot be a sea without the dry land, because without the land the sea has no limits and can provide itself with no definition. For that reason, the words "and the Lord saw that it was good" appear twice on the third day: once to mark the completion of the work of the second day, and once for the work of the third day, properly speaking.

And so once again, the sea: when was it made? In one sense, it had always just been there: "And the spirit of God hovered over the waters" (Gen. 1:1). Chronologically, it was brought into being on the second day. And yet, in another sense, it did not exist until it was given a name on Day Three and clearly set off from all other things (Gen. 1:10). On Day Two God made an expanse, and it divided the water which was under the expanse from the water which was over the expanse (Gen. 1:7). The text seems to make no fundamental distinction between the two waters; they differ only by virtue of the expanse itself. The angry sea and the torrential rain, both present in our everyday experience, are themselves a part of that original chaos, despite that by giving them the name "seas," God placed them within bounds.

How shall we begin to grasp hold of these strangers—the sea, and the fish who can live so peacefully in it? Well, perhaps, by beginning with something more human; so let us consider the Philistines, who were referred to by other literature of the time as a *sea people*.

Of their origins we know little or nothing—or, indeed, perhaps we know too much. From Chapter Ten of Genesis, the great chapter on the generation of the peoples of the earth, we know that Ham's second son Egypt (according to most English translations)

> fathered Ludim, and Anamim, and Lehabim, and Naphtuhim, and Pathrusim, and Kasluhim, from whom the Philistines came, and Kaftorim. [Gen. 10:13–14]

Not only does this passage feel strange syntactically, it is also at odds with several sources in Scripture which all affirm that the Philistines came from Kaftor and not from Kasluhim:

> Did I not bring up Israel from the land of Egypt, and the Philistines from Kaftor and the Syrians from Kir? [Amos 9:7]
>
> For the Lord is destroying the Philistines, the remnant of the coastland of Kaftor. [Jer. 47:4]
>
> As for the Avvim, who lived in villages as far as Gaza [a Philistine town], the Kaftorim, who came from Kaftor, destroyed them and settled in their place. [Deut. 2:23]

These verses, along with the rather ungainly character of the word order of our verse, suggest that the word *Kaftorim* has fallen out of place, and that the passage should instead read:

> Egypt fathered Ludim, and Anamim, and Lehabim, and Naphtuhim, and Pathrusim, and Kasluhim, and Kaftorim, out of whom the Philistines came.

The phrase *came out of* (יָצְאוּ מִשָּׁם) occurs nowhere else in the chapter and clearly marks a break in the rhythm of phrases like "and the sons of ... were ...," which occur no less than eleven times in the course of the chapter as a whole. Now while the word "to come out of" can occur in a phrase like that in Exodus 1:5, "to come out of the loins of Jacob" (יֹצְאֵי יֶרֶךְ־יַעֲקֹב), when it is used by itself it does not imply family origin.

Moreover, the word מִשָּׁם (misham) in Genesis 10:14 does not mean *from whom,* as most translators render it, but rather *from where*. This makes the final translation read:

> Egypt fathered Ludim, and Anamim, and Lehabim, and Naftuhim, and Pathrusim, and Kasluhim, and Kaftorim, from where the Philistines came.

We do not know how the Philistines happened to be in Kaftor, nor are we told who their fathers were. We do know that Kaftor is the island of Crete[2] and that the Philistines are of unknown origin and friends to a giant whose lineage can be traced back to the time before the flood. They were an island people lately come out of the Western Sea, and living in part of what had been the Promised Land.

But let us assume for the moment that the Philistines did in fact come from Ham—Ham who had the misfortune of seeing his drunken father naked and asleep in the tent.[3] As we read the book of Genesis, we are forced to participate in Noah's drunken stupor. The wine of oblivion shapes the whole of things from

2 See "Who is a Hebrew?" page 3 above.

3 Gen. 9:22ff.

that point on in the book. The names Adam, Eve, Cain and Abel, Methuselah, Seth and Enoch, even the serpent and the Garden of Eden—not to mention the flood itself—none of them will ever be mentioned again within the Torah or the books of the Earlier Prophets.The world will be as though none of them had ever existed.

The origins of the whole must be stated in some form; but from the biblical point of view they must also be forgotten. They may not be harkened back to, either as a paradigm or as a way of understanding. The temporal beginnings must be completely superseded by a covenant; but that is exactly what Ham cannot do.

We are not told whether Ham intended to look or not. It doesn't matter; he has seen what he should not have seen and knows what he should not know. That is his curse. He cannot participate in the waters of forgetfulness. His mind had been so transfixed by the sight of those antediluvian origins (which lie behind any political society) that he could never trust any covenant. Since he could never join in any covenant, it followed that he was at least in part bound to become a "servant of servants" (Gen. 9:25). The fundamental mistrust that the sons of Ham had for any covenant may be best seen in Ham's grandson Nimrod, who built the tower to escape any future flood. He relied on nothing, trusted nothing. Using brick for stone and slime for mortar, he built it in a plain so as not to make use even of a mountain.[4]

The first Philistine that we actually meet is the king of Gerar, Abimelech, the man to whom Abraham introduced his wife Sarah by saying, "This is my sister." In his own way, Abimelech is a good man. But he is good only in his own private way, a way which cannot be taught, and hence one which cannot be passed on to or learned by others. His way is a kind of unsophistication.

Abraham's dissembling had nearly led to Abimelech's death:

> But God came to Abimelech in a dream by night and said to him, "Behold, you are a dead man because of the woman whom you have taken, for she is a man's wife." [Gen. 20:3]

Yet Abimelech's only reaction to what Abraham had done to him was to ask, naively, "What did you see, that you did this thing?" (Gen. 20:10). He naturally assumes that Abraham would not have acted as he did without believing himself to have good cause. Abimelech is totally guileless, but equally incapable of recognizing guile. Noble though he is, his followers (as we learn when we meet him for a second time) are all thieves, while he is completely

---

4 Gen. 10:1ff.

oblivious to everything.[5] Abimelech is indeed no king at all. Such is the life of the blessed fish.

Some years later we meet Abimelech for a third time (Gen. 26:1–10); but again trouble arises, this time between the men of Abimelech and the servants of Isaac. The Philistines had filled with earth all the wells which Abraham had dug.[6] A full appreciation of this episode would require a more complete understanding of the character of Isaac than can be presented within the confines of this lecture; for the present, however, there is little more to say about Isaac than that he is the son of Abraham and the father of Jacob. James Joyce was insightful when he wrote of a *bland old Isaac*. Isaac's whole life was arranged for him, first by Damascus Eliezer, and later by his wife Rebecca. He liked his venison, and was careless enough to be caught red-handed by Abimelech.[7] Nevertheless, everything he did always came out right.[8] His one great deed was to keep his father's wells open. In his sleep and blindness, he showed that traditions well laid could rest in the hands of a blind but constant keeper, to arise again, refreshed, in another generation. That was his blessing. He was the blind keeper.[9] Isaac had the one virtue the Philistines, men of the sea, lacked.

In this context we can see yet another way of understanding the relation between Israel and the Philistines. From the time the two first met on the field of battle, the Israelites were constantly referred to as "Hebrews" (I Sam. 13:3–7), a term regularly used only by Philistines.[10] Aside from its use in Genesis 14:13, which would require a very long explanation, it was first used of Joseph when he became a slave in Egypt.[11] The name was consistently applied to all the Sons of Israel while they were slaves; but once the Red Sea is crossed, the word "Hebrew" (when it is not being used by a Philistine) occurs only in the phrase "Hebrew slave."[12] From the Philistine point of view, the Israelites are nothing but slaves because, for Philistines, *law as such* is slavery. The Philistines, on the other hand, are the only people whom the Israelites call "the foreskinned,"[13] since they, above all men, are in their natural state and know no covenant.

According to the Book of Joshua, God's original plan for the Promised Land included the land recently occupied by the Philistines.[14] But when God saw that not all the people had the virtue of Isaac, He decreed that the Philistines would remain "to put Israel to the proof" (Judg. 3:1–3). Israel and the Philistines had

5 Gen. 21:25ff

6 Gen. 26:14–22

7 Gen. 26:8ff

8 Gen. 26:12–14, 22–32 et al.

9 Gen. 26:24

10 I Sam. 4:9, 13:19, 29:3

11 Gen. 39:17

12 Exod. 21:2 et al.

13 Judg. 14:3, 15:8; I Sam. 14:6 et al.

14 Josh. 13:3

arrived together, one from the east and one from the west, one from the desert and one from the sea. One brought law, the other brought chaos. There would be wars, but neither would finally conquer the other. Thus it became clear that Israel could not live without the Philistines! But why should that be—what possible need can Israel have of the sea?

Samson's encounter with the Philistines may not solve the problem, but may at least help to define it with greater clarity. By trickery, the Philistines were able to answer Samson's famous riddle about the honey he had found one day in the rotting carcass of an ass: "Out of the eater came forth the eatable, and out of the powerful came forth the sweet" (Judg. 14:14). But we must try to answer the riddle without any tricks. *How can honey come forth out of the beast?* That is, what kind of wisdom can come out of turbulent power? Samson's riddle is actually a form of our earlier question: What need can Israel have of the sea? The fourth chapter of the First Book of Samuel presents a glimpse into what might be called a provisional solution.

The sons of Eli, and with them the whole of the priesthood in Shilo, had grown corrupt.[15] Before the priesthood could be renewed, the ark would be captured and would remain for some time in the hands of the Philistines.[16] If we think about the importance of water in the cleansing of any defiled object, we can begin to see the fascination which the sea in all its deadly turbulence held for the biblical author. Unlike the peace and lawful order which God wished to establish among men, chaos, threatening as it is to human existence, may be the only thing in our visible universe that is not open to guile, spite, or corruption.

While the ark of God was in the hands of the Philistines, a great battle took place between it and the idol Dagon. The outcome was disastrous for the Philistines. God sent them a plague of hemorrhoids and mice, and "The head of Dagon and the palms of both his hands lay severed on the threshold, but Dagon remained upon it" (I Sam. 5:4). God did not destroy Dagon, but set him within his limits as he had done to the sea. In typical Philistine fashion, they carefully return the ark, sending along with it five golden hemorrhoids and five golden mice, again causing us to wonder about the relation between the beautiful and the ugly, or between "the sweet and the strong."

When the ark returned to Beth Shemesh it had somehow changed; but the people had not yet changed with it. Fifty thousand and seventy of them were struck by the Lord just for having looked at it.[17] All we know of these people is that they were indeed of the sons of Aaron and so entitled to hold the ark—yet

15 I Sam. 2:27ff

16 I Sam. 2:34, 4:4

17 I Sam. 6:19

they could not.[18] Almost without thinking, they called for the men of Kiriath-Jaarim; and Abinadab accepted the ark without any fear and kept it safely for twenty years.[19] Now it must be remembered that these men were not Israelites. They were some of the Gibeonites who had tricked Israel into signing a covenant with them back in the days of Joshua;[20] and the text makes it clear that even in Samuel's day the people were still sensitive to the difference.[21] Israel was still not prepared for the ark; and, except for the disastrous time King Saul (already out of favor with the Lord) took it into battle,[22] the ark remained in the house of Abinadab and thus out of the hands of the sons of Israel until the time of King David.[23] But that would take us well ahead of our story, and we must return to the time of Samuel.

The rise of kingship in Israel was a long and very complicated affair; but even before Saul had been confirmed, God made it the first duty of the king to save Israel from the Philistines.[24] The first time we see them, however, is under quite different circumstances. The Philistines were present when Saul was first sent by Samuel to be "among the prophets."[25] Now it must be remembered that at this point in the book, those who were called prophets were not the men we normally think of. They were men who came down from the infamous "high places" with harps, tambourines, flutes, and lyres. At best they were utterly useless;[26] at worst they conjured up notions akin to bohemianism, if not to madness itself.[27] The book even goes out of its way to mention that neither Moses nor Samuel themselves were called by the name "prophet" in their own day.[28] But this is not the time to tell the tale of how the book itself understands the metamorphosis which prophecy underwent and which ultimately led to the rise of men like Elijah and Elishah, and finally to men like Isaiah and Jeremiah.[29]

Samuel seems to have sent Saul forth to give him a new spirit; but when the phrase "Is Saul also among the Prophets?" occurs in the text for the final time we shall see clearly what the Philistines, who were quietly there from the beginning, may have already known for a very long time, that is, the close connection between Saul's time spent with the prophets and his eventual madness.[30]

18 Josh. 21:13–16

19 I Sam. 7:1–2a; II Sam. 6:3–4

20 Josh. 9:17, 6:4

21 I Sam. 7:2b

22 I Sam. 14:18

23 II Sam. 6:1ff

24 I Sam. 9:19

25 I Sam. 10:11

26 Judg. 6:8

27 Exod. 15:20; Num. 11:25; I Sam. 10:5

28 Num. 12:6; I Sam. 9:9

29 See my chapter "On Prophecy" in *The Lion and the Ass* (note 1 above).

30 I Sam. 10:11-12, 19–24

Saul handled himself well in his first battle, which was against the Ammonites;[31] but when it came to the Philistines, things were no longer the same. Fear caused him to attack without waiting for Samuel to perform the sacrifice;[32] for he was not able to share Jonathan's spirit that "there is no restraint upon the Lord to save by many or by few" (I Sam. 14:6). Although Saul was stripped of his kingdom that day, the war went on;[33] but it did not go well. In what seems to be another sign of his approaching madness, Saul commanded the people not to eat anything until after the battle.[34] It was Saul's mad attempt to rid himself of the Philistines within himself by denying all natural needs. The results were disastrous, as Jonathan understood from his taste of honey;[35] and among the people eating took its wildest form.[36]

After the war, Saul seems to have been offered something like a reprieve if he could destroy the Amalekites; but he could not bring himself to kill Agag, their king.[37] The relationship between the Amalekites and the Philistines is a tale too long to be told here; but suffice it to say that Amalek, as the grandson of Esau, was Israel's closest relative.[38] We assert here without argument that, as the Philistines are the *sea without*, so Amalek is the *sea within*. If Saul could conquer that, he would indeed be king; but he could not. Samuel loved Saul, but Saul was mad and madmen cannot be kings.[39]

An evil spirit was upon Saul and he sent for "a man of music"[40] to cure him of his madness. Music—song without speech—a strange concept in Israel! It is the first time the word "to play" (*nagan*) comes up in the Bible, and it will appear only one more time in the set of books from Genesis through Kings. Indeed, *music* does not seem to be part of the *way of law* at all. We must keep a good eye open for the many kinds of music which affect Saul, and the many ways in which he is affected by them. On this occasion it refreshed him and caused the evil spirit to depart.[41] But this "man" who charmed away the king's madness was in fact just a boy, the boy David, a ruddy, fine-eyed young boy[42]—but a boy who, we know, had already been appointed king.[43]

David was perhaps a bit impetuous; and one might well suspect that his brother Eliab was nearly right when he said, "I know thy insolence, and

31 I Sam. 11:5
32 I Sam. 13:11
33 I Sam. 13:13
34 I Sam. 14:24
35 I Sam. 14:27
36 I Sam. 14:24–45
37 I Sam. 15:1–33
38 Gen. 36:12
39 I Sam. 15:33–34
40 I Sam. 16:15
41 I Sam. 16:23
42 I Sam. 16:12
43 I Sam. 16:13

thy peevishness, for thou art come down to see the battle" (I Sam. 17:28). Nonetheless, it was there that "he smote the Philistine, and killed him, but no sword was there in David's hand" (I Sam. 17:50). One cannot help comparing David's daring to Saul's fear on first meeting the Philistines, especially when one remembers that "Neither sword nor spear were found in the hand of any of the people that were with Saul and Jonathan: but with Saul and with Jonathan his son were there found" (I Sam. 13:22).

Saul was pleased with David at first, but then then came that other song, the women's song,

> Saul has slain his fifty score, David has his ten times more, [I Sam. 18:7]

over and over, again and again, with a sing-song meter that the rendition given here has attempted to capture. The Hebrew rings with a doggerel kind of rhyme, and Saul's evil spirit returned.[44] This time even David's song could not help.

Saul's plan for getting rid of David included marrying him off to one of his daughters, either Meribah or Michal. It didn't matter which; after all, they were both his daughters. "But Michal loved David" (I Sam. 18:18–20). That was the thing Saul could not understand: Michal and Jonathan both turning against him and being in love with David. What was this fascination that was stronger than family, stronger than law? As Saul later put it to Jonathan, "Thou perverse and rebellious son, do I not know that thou hast chosen the son of Jesse to thine own disgrace, and to the disgrace of thy mother's nakedness?" (I Sam. 20:30). Saul speaks these words, and we remember Noah.

After Saul's attempt on his life, David ran to Nayoth. Saul first sent messengers and ultimately came himself.[45] When he arrives we see the full implication of that question we had heard so many years before: "Is Saul also among the prophets?" The text reads:

> And he stripped off his clothes also and prophesied before Samuel in a like manner, and lay down naked all the day and all the night. Wherefore they say, "Is Saul also among the prophets?" [I Sam. 19:24]

Only now do we see the full implications of that first meeting between Samuel and Saul. The innocent boy is now a ruined old man lying naked on the ground, with a stern Samuel standing over him. Is this what that haunting phrase "Is Saul also among the prophets?" meant even from the beginning? One day Saul the innocent will have Doeg the Edomite slay Ahimelech and destroy the whole city of Nob, even when servants would not obey a king to "put forth their hand to fall upon the priest of the Lord" (I Sam. 23:17). What was it that the boy who went looking for his father's asses found? How has

44 I Sam. 18:10

45 I Sam. 19:20–22

trust of family and fear of the sea led him to the mistrust of family and to a maddened embrace of the sea?

David fled to Nob. There he took the bread of the priests and the sword of Goliath.[46] By breaking the law ever so gently and taking up the Philistine weapon, David began to enter the sea. At first he went too quickly and was rejected by the servants of Achish. But David could learn to feign the madness which held Saul tight, and so escaped.[47] But how would David continue his journey back to the sea?

When David left Achish, he went first to Adullam and then to Moab. It was a very long journey and he had to, in effect, travel back over many centuries. The text reads:

> David therefore departed thence and escaped to the cave of Adullam; and when his brethren and all his father's house heard it, they went down thither to him. [I Sam. 22:1]

To see what the cave of Adullam meant to "all his father's house," we must go back to the days of his old father, Judah.

◆

It was the time that Joseph had been sold as a slave into Egypt. Reuben was the eldest and felt the responsibility which that implied. It was his idea to put Joseph in the pit, thinking to come back later and return him to his father.[48] But the oldest is not always the wisest. It was Judah who rightly saw the depth of the problem. It would not just go away, as Reuben had thought; no, the boy Joseph had to leave. But how to manage it? Judah's plan was twofold and it was masterful:

> And Judah said unto his brethren, "What profit is it if we slay our brother and conceal his blood? Come, and let us sell him to the Ishmaelites, and let not our own hand be upon him; for he is our brother and our flesh." And his brothers were content. [Gen. 37:26–27]

It was only after he had shown them that they had nothing to gain by killing their brother that Judah appealed to any natural abhorrence they might have of fratricide. Without an appeal to the lower, they would not have heard an appeal to the higher. Without the appeal to the higher they would have learned nothing.

And so Joseph was sold into Egypt. His brothers killed a kid and dipped Joseph's coat into its blood; and taking it to their father they said, "Please to recognize whether this be thy son's coat or no" (Gen. 37:31–32).

---

46 I Sam. 21:2–10

47 I Sam. 21:11–16

48 Gen. 37:21–22

It was about this time that Judah went to lodge with Hirah the Adullamite. He had left his brothers, who were no brothers at all since they did not know what it meant to be "of our flesh." He intended to make a fresh start in a new country, but things did not go well.[49]

Judah had three sons: Er, Onan, and the youngest, Shelah. In time he took a wife for Er: her name was Tamar. "But Er was wicked in the sight of the Lord, and the Lord put him to death" (Gen. 38:7); we are not told what was Er's offence. Onan was then obligated to raise up offsprinng for Er, but he refused and spilled his seed upon the ground, for which he too was slain by God. Tamar returned home and waited for Shelah to grow up. But when the time for Shelah's marriage to Tamar arrived, Judah—fearing that Tamar brought death to her husbands, put off the match.

When Tamar heard one day that Judah had gone sheepshearing with his friend Hirah the Adullamite, she dressed as a harlot and stood by the road. Upon Judah's return, he lay with the harlot; and when they parted, he left her his signet and his cord and his staff as pledges for payment. But when he sent Hirah the Adullamite to redeem the pledges, no harlot was to be found.[50] Some time later, Judah was told that Tamar was pregnant, and he went to stone her as a whore. But she confronted him with his signet, cord and staff and said to him, "Please to recognize..."[51]

It's hard to imagine that Judah even heard the rest of her speech. Those words must have cut more deeply than even Tamar could have understood. It had been so many years ago, and now he was hearing them all over again. Why must he go back to all that?

When we next see Judah, he is back among his brothers. They had all gone down to Egypt to buy grain, but suddenly found themselves standing in front of Joseph, accused of being spies. Suddenly, they all remembered:

> Then they said to one another, "In truth we are guilty concerning our brother, in that we saw the distress of his soul, when he begged us and we did not listen. That is why this distress has come upon us." [Gen. 42:21]

Reuben, the eldest, spoke: "Did I not say unto you, do not sin against the boy? But you did not hear; and now his blood is required" (Gen. 42:22). Reuben felt the problem, but he began to pull away from his brothers just as Judah was returning to them. Reuben was insensitive to the brothers' recognition of their own guilt and seemed only to wish somehow to exclude himself from their confession. But, for the moment, Judah remained silent.

---

49 The following incidents in the life of Judah were related with a different focus in "Judah: Seeds of Kingship," above.

51 Gen. 38:25

When they returned to their father Jacob, he, now mourning the loss of both Joseph and Simeon, would not allow Benjamin to be taken back down into Egypt. Reuben again spoke up: "Slay my two sons, if I bring him not to thee." (Gen. 42:37). This contorted show of concern, however, was far from what Jacob could possibly have wanted; and Reuben's suggestion was simply ignored. And Judah still waited. But when he judged the time to be right, Judah spoke to his father simply and soberly:

> "Send the boy with me, and we will arise and go, that we may live and not die, both we and you and also our little ones. I will be a pledge of his safety. From my hand you shall require him. If I do not bring him back to you and set him before you, then let me bear the blame forever. If we had not delayed, we would now have returned twice." [Gen. 43:3–10]

This time everything was presented simply; nothing was in excess, and Jacob agreed.

By the time they arrive in Egypt, Judah has clearly emerged as leader. He recounts the past; in almost childlike terms he quietly but firmly tells the Great Magician of all the pain he has caused Jacob. When their discussion is over, the reader knows that sobriety is more powerful than magic; and that the "Ephrati" who will one day be king will not be a man from the tribe of Ephriam, the son of Joseph, but a man from the city of Ephrath, that is to say, from Bethlehem.[52]

How much did Judah know when he privately and discreetly went up to Joseph] and said,

> "O, my lord, please let your servant speak a word in my lord's ears, and let not your anger burn against your servant, for you are like Pharaoh himself. My lord asked his servants, saying, 'Have you a father, or a brother?' And we said to my lord, 'We have a father, an old man, and a young brother, the child of his old age. His brother is dead, and he alone is left of his mother's children, and his father loves him.' Then you said to your servants, 'Bring him down to me, that I may set my eyes on him.' We said to my lord, 'The boy cannot leave his father, for if he should leave his father, his father would die.' Then you said to your servants, 'Unless your youngest brother comes down with you, you shall not see my face again.' [Gen. 44:18–23]

Did Judah know that he was accusing his brother of risking parricide?

> "For your servant became a pledge of safety for the boy to my father, saying, 'If I do not bring him back to you, then I shall bear the blame before my father all my life.' Now therefore, please let your servant remain

52 See "Judah: Seeds of Kingship," page 16 above, where the identity of the "Ephrati" is discussed at length.

instead of the boy as a servant to my lord, and let the boy go back with his brothers." [Gen. 44:32–33]

Tamar had asked, as Judah's pledge, his signet, his cord, and his staff—all signs of his identity. Now Judah is willing to become himself the pledge, giving up his identity by once again becoming a slave in Egypt, to restore the family he had broken.

Judah looked at Tamar and—finally—heard the rest of her speech. That was the story David could learn from the walls of the cave of Adullam.

◆

When David left Adullam he took his mother and father and brought them to the king of Moab.[53] Again he has gone back in time—back to the wonderful days of his mother Ruth[54] and to the horrible nights of his father Lot.[55] Have we then begun to see some kind of answer to the question of how David can continue his journey to the sea? That journey seems to require a going back, a knowledge of who he was, where he came from, and why it was he who had to be king. But having come so far, it seems we are faced with nothing but a knot of contradictions. How can David be a teacher where there is nothing to be taught? How can lasting tradition be established in the ever-changing sea? But there is even greater contradiction. How does David's blessing—his going back, his seeing behind—differ from the curse of Ham? No solution to this problem appears within our present horizon, and we must proceed with great care.

After David left Moab, the prophet Gad told him to go to the land of Judah.[56] There he heard that the Philistines had attacked Keilah, and David asked God whether he should rescue them. The answer was "Go!" But because his men were afraid of the venture, David asked God a second time; and God, even more emphatically, said, "Arise, go!" God's implied promise of victory was true, but deceptive. The battle was a great success; but when David asked, "Will Saul come down against me?" God said, "He will come!" And when David pressed further, "Will the men of Keilah deliver me into the hand of Saul?" the laconic God answered, "He will deliver!" God knew that all the time! David had been tricked, and he was now a fugitive from both sides. To return to the sea, to be in the act of becoming a fish, means to be neither a man nor a fish. But even in that condition, David was able to make a covenant with one private

53 I Sam. 22:3

54 Ruth 4:22

55 Gen. 19:36–37

55 Gen. 19:36–37

56 I Sam. 22:5. The ensuing events are recounted in I Samuel 23:3–12.

man, Jonathan. It may have been the most important; by its means, somewhere between the land and the sea, Jonathan ceded the kingship to David.[57]

We are next presented with two battles. In many ways, for example in their names and incidents, they are quite repetitious. But the biblical author often uses repetition as a way of carefully examining a difference or change.

The first battle began when the Ziphites, an otherwise practically unknown people, told Saul in Gibeah that David was in Hachilah, south of Jeshimon. David heard about it and fled to a place called Maon, and Saul gave chase. On this occasion David would have lost the battle had it not been for the Philistines, who happened to attack Saul at the same time. There was no particular concern for David on their part; nonetheless, David was saved by the Philistines.[58]

David fled to En-Gedi, and Saul pursued. Saul went into a cave in order to relieve himself; but David happened to be hiding in the same cave. Rather than kill Saul, however, David cut his skirt. But even for that act he later felt remorse; and Saul was reconciled to him.[59]

The second battle started out as an exact repetition. Again, the Ziphites told Saul, who was in Gibeah, that David was in Hachilah south of Jeshimon. But then things began to change. This time it was David who attacked; and rather than being surprised in the cave, he planned his entrance carefully. And far from feeling remorse, he seriously, but also somewhat playfully, chided the king's men for guarding Saul so poorly. The most important difference, however, is that in the second affair David did not come nearly so close to the person of the king. Saul, though asleep at the time, was at least in a somewhat less delicate position then he was the first time. And while David had one of his men take the King's spear and cruse of water, he himself did not come near the king's person.[60]

We must note that while David took both the spear and the cruse of water, and even mentions both of them when chiding the king's men, he returned only the spear.[61] We should also note that the word for "cruse" is rarely used elsewhere in the Bible, and then only in connection with the mystical acts of Elijah.[62] To understand this we must consider the story which the biblical author chooses to tell in the interlude between the two wars, during which David hears what he himself called "Abigail's advice" (I Kings 25:33) to "shed no blood causelessly" (I Kings 25:31). Saul the man need not die; but he could no longer be king. He could no longer hold that mystical cruse, because his

57 I Sam. 23:16–18

58 I Sam. 23:16–28

59 I Sam. 24:1–22

60 I Sam. 26:1–25

61 I Sam. 26:22

62 I Kings 17:12–16, 19:6

inability to face the sea without going mad rendered him incapable of keeping the waters of chaos in confinement.

Abigail's advice to "shed no blood causelessly" was all that David had. David was now prepared to enter into the depths of the sea to meet the Philistine. But would he ever find the long road back? Can a man kill a monster, even to protect humanity, without becoming a monster himself? God's Law, the one God had given to Moses on Mount Sinai, said "Thou shalt not kill"; but David would kill. Some would render it "Thou shalt not murder"; but murder presupposes civil law, and David was going into the sea. Abigail's advice will be a turning point in the life of David. Weak as it may sound, it will hold true even in the roughest sea, and it will bring him back to the land and to the law. We know from the story of Uriah and Bath-Shibah that the biblical author is fully aware that the way is not easy to find. We also know that one day David will find his way home; but by then his son, the first one he had with Bath-Shibah, will be dead and David will have suffered much. The Bible seems to be warning us that some never make it back home. But we must return to our story.

After David had spoken with Abigail, he prepared to become a fish; and Achish accepted him without question. Achish reminds the reader of Abimelech. He too is good, kind and generous,[63] and completely taken in by David, who spent most of his time in his own all-important war against Amalek.

But for Saul things went badly. He became a broken man, with no one to turn to but the witch at En-dor. The reader knows that, as one who participates in the covenant, Saul was under personal obligation to stone every witch he encountered;[64] and now he was to meet one for the first time. Under protest, she raised Samuel from the dead. But Samuel rejected Saul again; Saul was mad and madmen cannot be kings. Yet the witch showed him compassion. In a manner that reminds the reader of the meal Abraham served the three angels, she promised little but served him a fatted calf.[65] For Samuel, Saul was a king who had to be rejected; but the witch, precisely because she had placed herself completely outside the law, could regard him as a man and pity him.

However, let us not become mere dreamers. One day, during the reign of King Manasseh, the author will be forced to remind us how deleterious witchcraft is to political life.[66] But to the extent that he has caused us to fall in love with the witch that we have been commanded to put to death, the author has, for the moment at least, caused us all to peer beyond ourselves, to become fish. The text has led the reader to acknowledge that while he must follow the ways

63 I Kings 27:6, 29:2–11

64 Lev. 20:27

65 I Sam. 28:22–25; compare Gen. 18:5–8

66 II Kings 21:7

67 Job 42:5

of Israel, there are others like Abimelech, like Achish, and even the witch, who follow other ways, and who are no less pleasing in the sight of the Lord. In this case, then, to *love one's enemy* does not mean to love him in his potential sameness, but in his actual otherness.

In the few remaining moments we could go on to see how David was able to restore the ark to Israel and bring Israel back to the ark. Instead however, let us consider a man who may have peered into that world beyond our world even more profoundly than did David; that is Job.

Job was perplexed. He wanted to believe the stories of the wise—that the good prosper while the bad sink and suffer; but all around him wise men suffered and good men were brought low. He had only to look at himself to know that not all men who suffer have evil hidden in their hearts. There were times when to proclaim such a doctrine serves only to increase an already undeserved pain. God was just; yet the salutary myth was not always salutary. Indeed, it could be cruel.

Then, in the tempest, God showed him many things; and Job was crushed. "I am," he said, "of little worth, and what shall I answer thee?" (Job 40:3). Most translators and commentators seem to be satisfied with Job at this point, and they treat the next three chapters as little more than repetition. But, as we shall see, God was not looking for recantation. He was demanding much more. Indeed, God meant it when He said "Gird thy loins like a hero!" (Job 40:7). For it will indeed take a hero. God wanted Job to see Him with the *seeing of his eyes* rather than hearing of Him with the *hearing of his ears*.[67]

That meant going beyond the world of men who speak, going into the watery realm of the great fish, the Leviathan. Job saw the beauties and the horrors of that realm. He saw the wild ass snorting at civilization, and the silly ostrich. He saw the grass which grew where no man was; and he saw the great Leviathan itself. There he saw God at play. Job seems almost to have thought about Samson and his riddle—how can the sweet come forth out of the strong? His answer has been deeply misunderstood.

The men of King James say: "Wherefore I abhor myself and repent in dust and ashes." The new Jewish Publication Society translation says: "Therefore I recant and relent, being but dust and ashes" (Job 42:6).

King James is right about the verb. The word indeed means "to abhor"; though perhaps it would be somewhat more accurate to say "to have contempt for." But the verb "to have contempt for" must have an object; and there is no word for "myself" in the text. Now if King James were right in what follows one would expect the *athnach*[68] mark to fall on the word they translate "abhor";

68 The Hebrew *athnach* (֑) indicates the major syntactical division in a biblical verse.

but it actually falls on the word they translate "repent." Moreover, that word by itself does not mean either "repent" or "relent" but to feel a deep sense of inner sorrow. In English, if one feels sorrow for some past action of one's own, one may be said to repent; but, as in English, the Hebrew verb "to sorrow" by itself implies neither guilt nor repentance.

Then too, the elliptical phrase "in dust and ashes" (עַל-עָפָר וָאֵפֶר), meaning "while sitting amidst dust and ashes," is a construction foreign to Hebrew syntax. On the other hand, it is quite normal for the word translated "in" (עַל) to follow the word for sorrow; and together they simply mean "to feel a deep compassion concerning...."

The words "dust and ashes" occur in two other passages in the Bible. In one place Job himself is speaking; and in the other, Abraham uses the phrase while playing the part of Job before God prior to the destruction of Sodom and Gomorrah.[69] In both cases the phrase means *man*, lowly man.

Job's last words, then, should be rendered, "I have contempt and compassion for mankind." Job had entered into a strange land. It was bigger, more terrifying, and also more tender, than any man had ever seen. But it was a world in which man is no more than a fleck, and his boils no more than a fly buzzing around his head. When Job prayed for the wellbeing of the comforters who had so tormented him, he showed that he had learned to live at ease in both worlds at once. He had lived in the waters and had returned to live on the dry land. The compatibility of contempt and compassion seems to be Job's answer both to Samson's riddle, and to Abigail's advice to David. Perhaps only a man who felt as Job did after God's first discourse could have understood the things that were shown to him in the second discourse. It was that which allowed Job and David to return to where Ham could not. One need, however, only consider the story of Bath-Shibah to see that the biblical author knows that the problems are vast and that he is not in possession of any magical solution.

NOTE: I am grateful to a student of mine, Mr. Holsted, for having shown me that Jonah was another character who *went beyond*; but Jonah learned only contempt, without compassion. Isaiah was still another who went beyond and saw the great fish. Now since, for him, the day will come when God will slay the monster (Isaiah 27:1), there is no contempt; there is only compassion. But for Job, God is not like an angry child who would smash the head of its toy (Psalm 74:13–14).

69 Job 30:19; Gen. 18:2

The Perversion at Peor

I

Let us begin by considerinng two biblical verses. They are both from the book of Deuteronomy, though there are many like them throughout the pages of the Bible. They are often thought to be contradictory. The aim of this discussion is to clarify the difference between them and to consider the relationship they have to one another.

> Fathers will not be put to death because of their children, nor will children be put to death because of their fathers. Each one will be put to death for his own sin (חֵטְא, *ḥet*). [Deut. 24:16]

and

> I the Lord your God am a jealous God, visiting the iniquity (עָוֹן, *avon*) of the fathers upon the children to the third and fourth generation of those who hate me. [Deut. 5:9]

The two verses are not speaking of the same fathers. The first verse relates to a single individual who has committed a *ḥet* worthy of death. He and he alone can pay the price. No one is to be put to death in retribution for the act of his father. In contrast, the second verse is speaking of the fathers of the past, whose *avon* has been so tightly woven into the fabric of society that all suffer, guilty and innocent. In our own country, the institution of black slavery comes to mind.

The word חֵטְא (*ḥet*), ordinarily translated as *sin*, was at first a term used among archers, meaning *to miss a mark*. The play on words was sufficiently fresh to allow Jeremiah to write:

> All you who bend the bow; shoot at her, spare no arrows, for she has sinned (חָטָאָה) against the Lord. [Jeremiah 50:14]

The word עָוֹן (*avon*), on the other hand, comes from the root עוה, which means *to twist*. Often it is translated *iniquity*. *Perversion* would be a more literal rendering. Isaiah writes:

> Behold, the Lord will empty the earth and make it desolate; and he will twist (וְעִוָּה) its surface and scatter its inhabitants. [Isaiah 24:1]

These origins suggest that if one has *missed the mark*, it would be best to try again; but if something has become *twisted*, trying again may not help.

As in English, the Hebrew word for *sin*, חֵטְא, can be either a verb or a noun. As a verb, it quite often has a plural subject:

> And I looked, and behold, you [Israel] had sinned (חֲטָאתֶם) against the Lord your God. You had made yourselves a golden calf. You had turned aside quickly from the way that the Lord had commanded you. [Deut. 9:16]

But as a noun, a *sin* is a particular sin of a particular person, either as stated by name or as posed in a general law:

> Then the chief cupbearer said to Pharaoh, I remember my sin (חֲטָאַי) today. [Gen. 41:9]

or

> Our father died in the wilderness. He was not among the company of those who gathered themselves together against the Lord in the company of Korah, but died in his own sin (בְחֶטְאוֹ). And he had no sons. [Num. 27:3]

And it can appear as a singular in the formulation of a law:

> You will not hate your brother in your heart, but you will reason frankly with your neighbor, lest you incur sin (חֵטְא) because of him. [Lev. 19:17]

Even though the next two quotations concern *the people of Israel*, they seem to be addressed to each individual rather than to the people as a whole:

> And speak to the people of Israel, saying, whoever curses his God will bear his sin (חֵטְא). [Lev. 24:15]

> So that the people of Israel do not come near the tent of meeting, lest they bear sin (חֵטְא) and die. [Num. 18:22]

The next passage is of special interest to us, because insofar as it refers solely to Jeroboam, "sin" is a noun; but when it also refers to Israel, it is a verb:

> But Jehu did not turn aside from the sin (חֵטְא) of Jeroboam the son of Nebat, which he made Israel to sin (הֶחֱטִא)—the golden calves that were in Bethel and in Dan. [II Kings 10:29]

It is difficult to assign a precise meaning to either word. עָוֹן seems sometimes to demand to be translated *guilt*, while other contexts require *punishment*. While a similar ambiguity is present in the word חֵטְא, our present discussion will confine itself to the consideration of עָוֹן.

The first to utter the word עָוֹן was Cain:

> Cain said to the Lord, my *avon* (עָוֹן) is greater than I can bear. [Gen. 4:13]

Cain, brother of Abel, was the first one to commit fratricide and also the founder of the first city. He was told to become a wanderer and to live the life that his brother, the shepherd, had been pleased to lead. Cain was to wander through the whole while not claiming any part of the whole as his own. But Cain did not wander. His *punishment* was never enforced. His *guilt* was the fratricide itself, and it all lay behind; now that Abel was gone, Cain could found a city by claiming part of the whole to be his own.

The term *avon* appears with a plural subject when Judah and his brothers, accused of theft by "the man" in Egypt, senses a connection between their present plight and the time when they—who would become the founders of the tribes of Israel—were about to commit fratricide by killing their brother Joseph:

> And Judah said, What shall we say to my lord? What shall we speak? Or how can we clear ourselves? God has found out the *avon* of your servants; behold, we are my lord's servants, both we and he also in whose hand the cup has been found. [Genesis 44:16]

In contrast to *ḥet*, a city can have an *avon*:

> As morning dawned, the angels urged Lot, saying, Up! Take your wife and your two daughters who are here, lest you be swept away in the *avon* of the city. [Genesis 19:15]

II

In the Book of Joshua we read:

> Have we not had enough of the *avon* at Peor, from which even yet we have not cleansed ourselves, and for which there came a plague upon the congregation of the Lord? [Josh. 22:17]

"The *avon* at Peor" refers to this episode:

> While Israel dwelled in Shittim, the people began to whore with the daughters of Moab. These invited the people to the sacrifices of their gods, and the people ate and bowed down to their gods. So Israel yoked himself to Baal of Peor. And the anger of the Lord was kindled against Israel. [Num. 25:1–3]

But what made it an *avon*? To address this question we must go back to the time when Balak the son of Zippor, king of Moab, "saw all that Israel had done to the Amorites":

> And Moab was in great dread of the people, because they were many. Moab was overcome with fear of the people of Israel. [Num. 22:2–3]

So Balak

> sent messengers to Balaam the son of Beor at Pethor, which is near the river in the land of his people to call him, saying, "Behold, a people has come out of Egypt. They cover the face of the earth, and they are dwelling opposite me." [Num. 22:5]

At this point we do not know who Balaam's people are. We only know that

Balak sent to Balaam, asking him to curse "the people come out of Egypt," because Balak had heard that

> He whom you [Balaam] bless is blessed, and he whom you curse is cursed. [Num. 22:6]

Balak's words sound very much like a mangled memory, tattered through the ages, of what God had once said to Abram:

> And I will bless those who bless you and him who dishonors you I will curse; and in you all the families of the earth shall be blessed. [Gen. 12:3]

Now according to Deuteronomy, Balaam son of Beor was from Pethor in Aram-Naharaim (Deut. 23:5); this is the land to which Abraham had sent his servant to find a wife for his son Isaac:

> The servant took ten of his master's camels and departed, taking all sorts of choice gifts from his master; and he arose and went to Aram-Naharaim to the city of Nahor. [Gen 24:10]

Thus Abram and Balaam, four hundred years apart, were both from Aram-Naharaim, though not from the same city. While Balaam was from Pethor, Abram was from Haran (Gen. 12:5); I do not think it is known how far apart the two cities were. Genesis also recounts that when Abram left Haran for Canaan, he took with him

> Sarai his wife, and Lot his brother's son, and all their possessions that they had gathered, and the people that they had acquired in Haran. [Gen. 12:5]

Notice that the company took *all* their possessions, but not necessarily *all* of "the people they had acquired." We are not told whether any of these people remained behind in Haran.

III

The next part of our inquiry begins with a verse from Exodus:

> I appeared to Abraham, to Isaac, and to Jacob, as God Almighty (אֵל שַׁדָּי), but by my Name the Lord (יי)[1] I did not make myself known to them. [Exod. 6:3]

Now God did indeed say to Abram:

> I am God Almighty (אֵל שַׁדָּי). Walk before me, and be blameless, [Gen. 17:1]

And as Jacob was leaving for Paddan-Aram, Isaac blessed him with these words:

> God Almighty (אֵל שַׁדָּי) bless you, and make you fruitful, and multiply you, that you may become a company of peoples. [Gen. 28:3]

---

1 Hebrew יי indicates the divine Name, not pronounced, and written out in full only in a sacred scroll.

But how did Isaac came by the words אֵל שַׁדַּי (*El Shaddai*)? Were they revealed to him by God, or had they become a tradition handed down by Abraham, father to son? We know only that at Luz, God said to Jacob, "I am God Almighty (אֵל שַׁדַּי). Be fruitful and multiply. A nation and a company of nations will come from you, and kings will come from your own body" (Gen. 35:11), reinforcing and adding to what Isaac had already said before.

Jacob's blessing to his sons, as they were about to go down into Egypt for the second time, was:

> May God Almighty (אֵל שַׁדַּי) grant you mercy before the man, and may he send back your other brother and Benjamin. And as for me, if I am bereaved of my children, I am bereaved. [Gen. 43:14]

Clearly the blessing had passed on into human hands. Jacob later said to Joseph:

> God Almighty (אֵל שַׁדַּי) appeared to me at Luz in the land of Canaan and blessed me. [Gen. 48:3]

Aside from our passage in Exodus, the words אֵל שַׁדַּי will never occur again in the whole of Tanakh; and thus when Jacob blessed Joseph he said:

> ... the Almighty (שַׁדַּי) who will bless you with blessings of heaven above, blessings of the deep that crouches beneath, blessings of the breasts and of the womb. [Gen. 49:25]

Nor will the single word שַׁדַּי ever appear again in Tanar,[2] except when Balaam speaks of himself as one "who hears the words of God, who sees the vision of the Almighty (שַׁדַּי,) falling down with his eyes uncovered" (Num. 24:4). He uses a word that had not been heard in many generations, as if he were following some older tradition going back to the days of Abram in Haran.

But now we must return to our passage in order to see what light it can shed upon Balaam.

> I appeared to Abraham, to Isaac, and to Jacob, as God Almighty (אֵל שַׁדַּי), but by my Name the Lord I did not make myself known to them. [Exod. 6:3]

On the surface this claim seems false, for Abram seems to have known the name of the Lord from the beginning:

> From there he [Abram] moved to the hill country on the east of Bethel and pitched his tent, with Bethel on the west and Ai on the east. And there he built an altar to the Lord and called upon the name of the Lord. [Gen. 12:8]

---

2 A Hebrew acronym denoting the biblical books from the Torah through the Early Prophets, plus Ruth.

However, let us look more closely at the Hebrew word for "make myself known," נוֹדַעְתִּי. It can be broken up into two parts: ודעתי and נ. One part contains the notion of *knowing*, plus what western grammar calls the first person singular. We, however, are interested in the other part; what is the effect of the נ?

Let us first look at a clearer case, the difference between לחץ and נלחץ, in the verses

> You shall not wrong a sojourner or oppress (לחץ) him, for you were sojourners in the land of Egypt. [Exod. 22:20]

and

> When the donkey saw the angel of the Lord, she smashed (נלחץ) against the wall and pressed Balaam's foot against the wall. So he struck her again. [Num. 22:25]

The action is virtually the same in both passages, but the נ tends to draw our attention to the subject rather than to the object. In the first example we think more about the effect of the action on the oppressed. In the second example the נ turns our attention back to the donkey. In most cases the נ form became equivalent to the passive voice, but in other cases it comes closer to being reflexive. Hence the translation of נוֹדַעְתִּי as "make myself known" in Exodus 6:3.

Abram had the *sound* of the divine Name, but it was just a sound without a meaning. God's self was not it. For Moses the Name was related to "I am" (Exodus 3:14), but Abram needed the intelligibility of אֵל שַׁדַּי.

All of this is of interest to us because Balaam, though he lived far from Egypt and at the time knew nothing of Israel, also called God by the divine Name, thus giving him one more connection to Abraham:

> So Balaam rose in the morning and said to the princes of Balak, "Go to your own land, for the Lord has refused to let me go with you." [Num. 22:13]

When Balak's messengers had delivered his request, Balaam replied to them:

> "Lodge here tonight, and I will bring back word to you, as the Lord speaks to me." So the princes of Moab stayed with Balaam. And God came to Balaam and said, "Who are these men with you?" [Num. 22:8–9]

We must remember that at this point Balaam has no knowledge of the people whom he has been asked to curse; and when God heard what the men were asking of Balaam, he said:

> You will not go with them. You will not curse the people, for they are blessed. [Num. 22:12]

But Balak was insistent:

> I will surely do you great honor, and whatever you say to me I will do. Come, curse this people for me. [Num. 22:15–17]

Balaam answered:

> Though Balak were to give me his house full of silver and gold, I could not go beyond the command of the Lord my God to do less or more. [Num. 22:18]

But evidently Balaam was not quite so steadfast as his brave speech would suggest, for he added:

> Please stay here tonight, that I may know what more the Lord will say to me. [Num. 22:19]

God then shaped the situation in a strange and apparently contradictory way. First he came to Balaam at night and said, "If the men have come to call you, rise, go with them; but only do what I tell you" (Num. 22:20). At the same time, however,

> God's anger was kindled because he went, and the angel of the Lord took his stand in the way as his adversary. Now he was riding on the donkey, and his two servants were with him. [Num. 22:22]

There follows the well-known story of the donkey who suddenly speaks. Seeing the angel standing in the way, the donkey turned aside three times, despite Balaam's angry blows. The third time resulted in Balaam's foot being squeezed against the wall—our previous example using נלחץ in Numbers 22:25:

> Then the Lord opened the mouth of the donkey, and she said to Balaam, "What have I done to you, that you have struck me these three times?" [Num. 22:28]

God's seeming contradiction came to its fruition when

> Balaam said to the angel of the Lord, "I have sinned (חָטָתִי), for I did not know that you stood in the road against me. Now therefore, if it is evil in your sight, I will turn back." And the angel of the Lord said to Balaam, "Go with the men, but speak only the word that I tell you." So Balaam went on with the princes of Balak. [Num. 22:34–35]

Balaam used the word חֵטְא (sin). That was the goal of God's contradiction and the story can go on. Balak went to Balaam and said:

> "Did I not send to you to call you? Why did you not come to me? Am I not able to honor you?" [Num. 22:37]

Balaam replied:

> "Behold, I have come to you! Have I now any power of my own to speak anything? The word that God puts in my mouth, that must I

> speak." Then Balaam went with Balak, and they came to Kiriath-huzoth. [Num. 22:38–39]

Perhaps because Balak thought that an act might be acceptable where speech was not,

> Balak sacrificed oxen and sheep and sent for Balaam and for the princes who were with him. [Num. 22:40]

Balaam accepted the plan and said to Balak:

> "Build for me here seven altars, and prepare for me here seven bulls and seven rams." [Num. 23:1]

Balaam then left Balak in order to return to the use of speech, saying:

> "Stand beside your burnt offering, and I will go. Perhaps the Lord will come to meet me, and whatever he shows me I will tell you." And he went to a bare height. [Num. 23:3]

But the nature of that speech had changed. In the past the Lord had spoken to Balaam. Now, however, the Lord *put a word in Balaam's mouth* (Num. 23:5), just as he had put speech into the mouth of the donkey. We first hear of this when Balaam says:

> From Aram Balak has brought me, the king of Moab from the eastern mountains, saying: "Come, curse Jacob for me, and come, denounce Israel!" [Num. 23:7]

Balaam's account is both true and not true. He was indeed asked to curse Jacob, but not by that name; so far as we can see in the text, Balaam never heard the names "Jacob" or "Israel" until he heard them coming out of his own mouth. At best he only knew that he was asked to curse "a people coming out of Egypt."

Balak now understands that God was not willing to destroy the whole of that people. Perhaps he thought that if the curse appeared less threatening, God might allow it:

> Then Balak said to him, "Please come with me to another place from which you may see them; you shall see only the outer part of them, and shall not see them all; curse them for me from there." [Num. 23:3]

But these are the words Balaam found placed in his mouth in answer to Balak:

> "God is not man, that he should lie, or a son of man, that he should change his mind. Has he said, and will not do it? Or has he spoken, and will not fulfill it?" [Num. 23:19]

By now Balak had all but given up and importuned Balaam neither to curse the people nor bless them. But Balaam reminded Balak that he had warned him from the beginning:

"Did I not tell you, all that the Lord says, that must I do?" [Num. 23:26]

Balak made one last attempt:

"Come now, I will take you to another place. Perhaps it will please God that you may curse them for me from there." [Num. 23:27]

But Balaam saw that "it will please God" not to curse, but to bless Israel (Num. 24:1), and Balaam declaimed:

"How lovely are your tents, O Jacob, your encampments, O Israel!" [Num. 24:5]

Balak turned furious. "You were to curse, and three times you have blessed! Get yourself out to your own place!" (Num. 24:1), to which Balaam replied:

"And now, behold, I am going to my people. Come, I will let you know what this people will do to your people in the latter days." [Num. 24:14]

When we first met Balaam, he was introduced to us as having come from "the land of his people." We never knew what people that was, nor perhaps did he. But now he speaks of "my people" and "this people." The next eight verses hint, or perhaps more than hint, that "this people," Israel, *is* his people. In accordance with Balaam's last words to Balak, these eight verses recount what was and foretell what will be. Much of what will come to pass does indeed manifest the glory that Balaam foresees; some of it does not.

Then Balaam rose and went back to his place. And Balak also went his way. [Num. 24:25]

Balaam has finally left Balak and has gone to his place. He has found his people, and we have no reason to think we will ever see him again. Immediately after Balaam's departure there follows the narrative we noted near the outset, the verses that begin to identify "the perversion at Peor":

While Israel dwelled (וַיֵּשֶׁב) in Shittim, the people began to whore with the daughters of Moab. These invited the people to the sacrifices of their gods, and the people ate and bowed down to their gods. [Num. 25:1–2]

We were not told how or why Israel came to be in Shittim, which is mentioned nowhere else in Tanakh, nor are we told how they came into contact with Moab. The use of the word וַיֵּשֶׁב, however, implies that they had set up a full camp. The next verse is the passage to which Joshua referred. It gives an account of the עָוֹן at Peor, an account we are now prepared to face.

So Israel yoked himself (וַיִּצָּמֶד) to Baal of Peor. And the anger of the Lord was kindled against Israel. [Num. 25: 3]

Israel had gone a-whoring many times before and would do so again, always returning in the end. But "yoking himself" seems to be quite a different matter. Yoking seems more intentional and perhaps more permanent than whoring.

Before going on it might be of some help to take a look at the only other two verses in which words from the root צמד occur.

> When they were at the great stone that is in Gibeon, Amasa came to meet them. Now Joab was wearing a soldier's garment, and over it was a belt with a sword in its sheath fastened (מְצֻמֶּדֶת) on his thigh, and as he went forward it fell out. [II Sam. 20:8]

and

> You give your mouth free rein for evil, and harness (תַּצְמִיד) your tongue for deceit. [Psalm 50:19]

As a result of Israel's having "yoked himself" to Baal of Peor,

> The Lord said to Moses, "Take all the chiefs of the people and hang them in the sun before the Lord, that the fierce anger of the Lord may turn away from Israel." And Moses said to the judges of Israel, "Each of you kill those of his men who have yoked themselves to Baal of Peor." [Num. 25:4–5]

Moreover,

> One of the people of Israel brought a Midianite woman into his family, in the sight of Moses and in the sight of the whole congregation of the people of Israel. [Num. 25:6]

Thereupon Phinehas, grandson of Aaron, grabbed a spear and killed both the man and the Midianite woman. His action stopped a punishing plague sent by God, but not before 24,000 people had succumbed to it (Num. 25:9). These included Zimri, a man of Israel and Cozbi, a Midianite woman, both of them children of their tribal chiefs.

> Then the Lord spoke to Moses, saying, "Harass the Midianites and strike them down, for they have harassed you with their wiles, with which they beguiled you in the matter of Peor, and in the matter of Cozbi, the daughter of the chief of Midian, their sister, who was killed on the day of the plague on account of Peor." [Num. 25:17–18]

This, then, is what Joshua called the perversion (עֲוֹן) at Peor.

> After the plague, the Lord said to Moses and to Eleazar the son of Aaron the priest, "Take a census of all the congregation of the people of Israel, from twenty years old and upward, by their fathers' houses, all in Israel who are able to go to war." [Num. 26:1–2]

The preparations for war require sixty-five verses, after which the narrative proceeds in an unrelated way, speaking about the daughters of Zelophehad, the question of leadership after the death of Moses, and about many kinds of offerings. In all, five suspenseful chapters go by until the war finally begins.

> The Lord spoke to Moses, saying, "Avenge the people of Israel on

> the Midianites. Afterward you shall be gathered to your people." [Num. 31:1–2]

The war was a total success:

> And they slew the kings of Midian along with the rest who were slain; Evi, Rekem, Zur, Hur, and Reba, the five kings of Midian, [Num. 31:8]

We are shocked and confused, however, by the last words of the verse:

> And they also killed Balaam the son of Beor with the sword. [Num. 31:8]

Why Balaam? His killing makes no sense—until we read:

> Behold, these [the Midianite women], on Balaam's advice, caused the people of Israel to act treacherously against the Lord in the incident of Peor, and so the plague came among the congregation of the Lord. [Num. 31:16]

Balaam has found his people; it is the people Israel. Balaam always knew he had a people, but it is doubtful whether he knew which people it was. Certainly we did not know; he was first introduced to us only as coming from "the land of his people." But how was it "his" people? Was it his to join or his to rule?

Recall how many circumstances in Balaam's life seem to echo Abraham's world. Balaam's reputation recalls the promise that God had made to Abram. Balaam and Abraham were both from Aram-Naharaim, though not from the same city; Abram's city was Haran, the place where he "acquired persons."

Balaam is the only person in the whole of Tanar to use the word שַׁדַּי (*Shaddai*) after the time of the Fathers, although for him the word אֵל (*El*) has somehow been lost. He is also the only one besides Abraham to refer to God by the divine Name prior to any contact with Moses.

In the first of the four speeches placed in Balaam's mouth it was said that he was asked to curse Jacob. That turned out not to be completely true; he was actually asked to curse some nameless people. He was never given the name *Jacob*; and it is by no means clear that the name would have meant anything to him, since he seems to have no knowledge of the name *Abraham* or the names of any of Abraham's descendants until God put them into his mouth in the four final speeches.

In light of all these echoes, I think it possible that Balaam was a descendant of one of the persons that Abraham acquired while in Haran. Such a possibility fits well in time and space—and with the changes in text that typically happen over time. Balaam had a word from Abraham's time that no one else had, but he did not have names of his own time until God placed them in his mouth.

Balaam came to join his people; and so he did, in the tightest bond of horror he could think of. Or have we quite misunderstood what Balaam meant when he spoke of "my people?" Perhaps he did not mean *the people I belong to* but rather *the people that belong to me.* Did he perhaps see himself as the legitimate successor to Abram and Abram's God? Like the false mother of the dead child in Solomon's court (I Kings 3:16–28) he felt, "If it isn't declared mine, let it die."

The Book of Job
Readings and Thoughts

> There once was a man of the land of Uz, and his name was Job; the man was whole-hearted and upright, one who feared God and shunned evil. And there were born unto him seven sons and three daughters. His possessions also were seven thousand sheep, and three thousand camels, and five hundred yoke of oxen, and five hundred she-asses, and a very great household; so that this man was the greatest of all the men of the east. [1:1-3]

It sounds like the beginning of some wonderful fairy tale, set in the mysterious land "of the east." In the Hebrew language the word for *east* can also mean *ancient*, and it conjures up the dream of a child's notion of wisdom and valor.

Our attention is caught by the mention of Job's seven sons and three daughters, seven thousand sheep and three thousand cattle. The numbers *seven* and *three* have always had a magic ring to them—whether because they are felt to possess some kind of perfection, or because they are odd and somehow unbalanced, I do not know. The childlike nature of the first and last sections of the Book of Job, both with regard to diction and to the use of repetition, gives them a kind of never-never-land feeling, especially when contrasted with the stark reality felt in the rest of the text. Turning from one to the other is like turning from Dick and Jane to Shakespeare. The central part of the book begins:

> May the day of my birth be lost in oblivion and with it that night in which it was said, "A hero (גֶּבֶר) has been conceived." [3:3]

The very being of a hero lies in being remembered; but for Job, this day and everything in it is to be forgotten. There is, then, a poignant contrast between the great joy felt by all on that day and the secret horrors it contained.

> "Why," says Job, "does He give light to those whom toil has consumed, or life to the bitter of soul?" [3:20]

Man has a certain light, an innate sense of what is just and what is unjust. For Job, no man can ignore that light as long as he finds it within himself; and yet it is in constant opposition to the manifest will of God as seen in daily events. Would we not be better off without that light? The world makes too much sense to make no sense, and yet it makes no sense. If Job had no reason, the world would no longer look unreasonable, and he could sleep more soundly.

Eliphaz speaks:

> How can one speak without being more than wearisome? But who can refrain from words? It was you who always disciplined and strengthened so many frail hands, you who had the words to pick up those that were stumbling and bolster the knees that were about to bend." [4:2–4]

It is of utmost importance to note the genuine good will with which Eliphaz begins to speak. Only in that way can we catch a glimpse of what it was that made loving friends turn against Job so brutally. For Job's friends, and for Job as well, the only proper home *for* man is the home *of* man—the home of man as it has been defined by the wisdom of the fathers. The only proper concern for man is his fellow man. Not to be at peace within that world is to be an outcast and a man of sin. Throughout all of what follows we must constantly remind ourselves that our daily lives depend upon such a world. Only in that way can we begin to understand why good men might turn brutal when that world is suddenly found to be under attack. But for Job that world has already begun to crack. Job deeply believes in a just God, and yet he has seen the just suffer in meaningless pain. The wise men have assured him that all will work out for the best, but it does not.

Eliphaz poses the central question of the book:

> Shall a man be more just than his God? [4:17]

But does the question even make sense? Apart from the will of God, is there any standard for justice in whose light God's actions can be inquired into? Even if there is, what is its relation to what we humans feel as being just and unjust? And if they should differ from that standard, what is the status of those human feelings? Job knows that this is a question which must be faced one day, but how and in what terms? He is not yet ready for it.

Eliphaz offers an assurance:

> Have no fear of the beasts of the earth, for you have a covenant with the rocks in the field, and the beasts of the fields will bring you peace." [5:22–23]

Covenant, *b'rit*, is the word used for Abraham's covenant with God; peace in this most fearful world is only achieved by a divine covenant with the rocks and the beasts. Threatening boulders are not held back in their places by any innate forces to be found within the rocks themselves, but by a divine covenant freely established by that God who answers those who come unto him.

But all of Job's questions presuppose the relevance of the way things appear to naked man. They imply taking seriously what we have called the surface of things; that fire goes up, that dogs bark, and that innocent men sometimes suffer. If arguments presuppose intelligibility, then Job's arguments presuppose the relevance of the way things are in themselves. In other words, they presuppose either something like *nature* in the ancient sense, or *laws of nature* in the modern sense.

It is hard to know to what extent or in whatever vague way Eliphaz could have been aware of such an alternative. However, it is clear that he knows that the notion of rocks and beasts obeying a covenant cuts deeply into Job's words.

> "Does not a mortal," says Job, "have a term of duty to serve here on earth, and are not his days like the days of a hired servant?" [7: 1]

◆

There is a strange and eerie, almost Kafkaesque, feeling to the next set of verses. No God is ever mentioned in them, and everything is stated in the indefinite third person plural or in the passive voice. Job presents man as feeling a horrible and meaningless, but absolute and almost sacred, duty to some nameless and totally unknown power. This sense of duty—this sense that there is something to which, and to which alone, a man must devote his entire life—Job feels these things too. Some may try to name it or endow it with intent and love, and they may be at peace with it; but for Job it has no name; it has no intent.

> Am I the sea or some monster," says Job, "that You set watch over me? When I said that my bed will show me compassion and my couch bear my complaint, You frightened me with dreams and terrified me with visions; and I preferred strangulation and death to my own substance. I have contempt; I will not live forever. Let me be, for my days are but the mist of a breath." [7:12–16]

For Job the central idea which holds human society together, that man is ever under the care and watchful eye of his maker, is crippling and ultimately fatal to the human spirit. The feeling of being watched, and therefore of being some kind of monster in need of being watched, has so completely overwhelmed Job's inner world that even in his dreams he is tortured by an amorphous sense of guilt that arises out of the sense of being watched. Job's need to contact the outside world of his three friends is so great that its watching God has become the author of his dreams.

Job continues:

> What is a mortal that thou shouldst magnify him? And that thou shouldst set thine heart upon him? Yes, and inspect him every morning and test him every minute. When will you let me be? You will not even let me alone to swallow my own spit. Supposing I have sinned, what have I done to you, O thou great watcher of man? [7: 17–20]

Again, as so many times before in the text, Job seems to have in mind a variety of psalm-like verses which must have been on everyone's lips; but for

Job they take on the cast of ironic horror. To be constantly watched, and hence never to be one's self for one's self alone is, for Job, to be less than human. To be watched as a thing out of place is already to be out of place—or else to be a thing that cannot know its own place but must be watched and kept in, like the sea. Being permeated by God both in mind and in body, he feels untrusted, and hence untrustworthy.

Next, Bildad begins to speak:

> Only ask of the first generations. Seat yourself firmly upon what their fathers had searched out; for we are only of a yesterday and know nothing, our days are but a shadow passing over the land. Will they not teach you and speak to you as the words come tumbling out of their heart? Can papyrus grow where there is no marsh? or can a reed flourish without water? While yet in their tender days, they wither before any grass, still unpicked. [8:8–12]

Bildad's argument seems to be that wisdom is not available to the human mind outside the context of a human, and hence a political, tradition reaching back to the fathers. The span of a single lifetime is too short to gather the experience or the insight which would be needed even to approach a way of life dedicated to an autonomous inquiry into the surface of things—such a life as that which a Socrates might one day lead.

The combined wisdom of the fathers who, over many ages, have slowly planted our roots by living through life, is to be trusted beyond the inquiries of a single man who must hold himself back from life in order to question it—no matter how thoughtful that attempt may have been.

Bildad seems to have compassion and even a kind of love for the tender reed who goes it alone—that is, for the man who does *not* "seat himself firmly in the ways of fathers" or nourish himself in the waters of tradition but instead tries to search out wisdom for himself. But much as Bildad may love such a reed, he sees it as a thing that cannot last. Other plants may be out there that can stand without the marsh, but not man, the tender reed. That man has forgotten God, and is lost.

> Then Job answered and said, "Yes, all that I know; but then what can make a mortal's justice apparent to God? [9:2]

Job now thinks that Bildad may be right: the distinction between the thoughtful and the thoughtless may not be visible from the highest point of view. The surface of things, the place where dogs bark, where fire goes up, and where good people suffer pain needlessly and die—from the highest point of view, all that may completely disappear. On the one hand, this conclusion leaves him confused and perhaps frightened. On the other hand, this confusion will at length lead Job to peer into a strange world well beyond his own, one which none of us has ever seen.

His words, "Though I am just my own mouth would condemn me" (9:20), seem to be somewhere near the heart of things as they have been ever since the world began to fall out of focus. Job knows that he is guilty in the only world that he ever knew, and yet he knows that he is not guilty. But Job's sociality, his need for human companions and fellowship, is so great that when the friends condemn and reject him, he takes the only course left open to him: he rejoins society by joining them in his own self condemnation.

When facing the world around him and the conditions it has placed upon his being, Job feels out of place, contorted, and perverse; and yet, since he knows of no crime or guilt, he knows of no way of expiation. He even begins to feel guilty of the sin of having seen his own innocence. Divine wisdom and everyday justice; each seems to mock and jeer at the other till the whole turns meaningless.

Job concludes by saying:

> My days have passed by. My ambitions have been snapped, all that my heart possesses. They claim it is day when it is night, and in the face of darkness they say that light is near. If I must take the Pit to be my home and spread out my couch in darkness, call out to the muck, "Thou art my Father" and call out "Mother" and "Sister" to the maggots, where then is my hope? Oh my hopes, who will ever take note of them? They have all sunk down into the Pit, and together they lie in the dust. [17:11–15]

We do not yet know exactly what were Job's hopes for himself and perhaps for the whole of mankind. We only know that they are incompatible with the notion that man is a maggot. The best we can do for now is to say that the notion of man as a worm seems to constitute a total denial of the ultimate relevance of the surface world as it appears to man.

Man is both a social and a rational animal. The two are not identical. Job, so far as he knew, was the first man to feel the full implication of their distinctness; and being the first, he met it unprepared while it was yet naked and unmediated by prior human thought.

We are born as social animals, weak, feeble, and in need of others. The every-day life of action requires well-defined horizons established either by the fathers or divine revelation, or by those ghosts of former thoughts that once lived in the minds of others and which we call our common heritage. Without them there would be insufficient grounds for communal action or communal life. But the need for clarity, and hence for autonomous understanding, does not believe in ghosts. It can only grasp them by bringing them back to life as a rethought thought and hence a reexamined thought. Sociality, on the other hand, knows that it can only live by learning to live with the ghosts; our lives depend upon it.

Well we remember Bildad's words:

Only ask of the first generation... [8:8]

But is that "asking" the asking of *piety and belief*, or the asking of *wonder and of doubt*? For Bildad the answer is clearly the former, and to indulge in the latter is not only to try to answer the riddle of the Sphinx, it is to threaten Laius as well. Reason and dedication to the surface, then, demand the one thing forbidden by divine law, tradition and sociality: that is, *autonomous understanding*.

But Job is both a rational animal and a social animal, and he is both in the deepest sense of each. He therefore lives under two shaggy horizons. For each, the other both is and is not. As we have seen, sociality can only articulate itself to itself as the forbidder, and thereby as also the rewarder or the savior; and this is the form in which Job is aware of it. This act of self-forbidding of what the self wants most is felt as guilt.

Chapter Twenty Five, the shortest by far in the whole book, which leads to abrupt cessation of all discussion between Job and his friends reads as follows:

> Then Bildad the Shuhite answered and said: Dominion and fear are His. He makes peace in His high place. Is there any number to His troops? Upon whom does His light not fall? How can a mortal think himself just before God, or what can clean anyone born of woman? Look high as the moon, nothing shines. Even the stars are not pure in His sight. And now what of these mortals, the maggots, or the son of man, the worm. [25:1–6]

Bildad ends with the words, "And now what of these mortals, the maggots, or the son of man, the worm?" We remember them well; back in Chapter Seventeen, Job had said, "If I must call out Mother and Sister to the maggots, where then is my hope?"

Job had always known that the claim, "Man is a maggot," would end conversation, because it denies the relevance of what, for Job, is the ultimate ground of any human conversation. If man is a maggot, the surface view of things, as it reveals itself to human thought, is of no ultimate relevance; for the surface is the surface precisely because it is a surface *for man*. It has no other being than to be the beginning for man. In spite of the constant strife within Job's soul caused by the contradictory claims of *the surface world* and *the fear of the Lord* to be the beginning of wisdom, what is first for man as such must be the beginning of human conversation.

Conversation, then, must come to an end precisely because Job has no proof that he is not a maggot. If the surface world made perfect sense, he would know that he was not a maggot; but it does not make perfect sense. It is full of enigmas and contradictions. It makes too much sense to make no sense, and yet it makes no sense.

And now, what of Bildad? He was always the most understanding, but now, he is the one to make the final break. We ask ourselves why. Perhaps the first thing to notice is that, while he has totally abandoned any attempt to come to terms with Job, Bildad has never condemned Job personally. He seems to have learned from Job that any attempt to find a compromise on the question of the status of the surface world would be meaningless. But recall his words, "Look high as the moon, nothing shines. Even the stars are not pure in His sight." Bildad has taken the other path: nothing shines and man is a maggot! Bildad has seen that the only viable alternative to Job's position requires diminishing the status of both man and the visible universe.

To put it otherwise, Bildad back in Chapter Eight took the position that no single man, by himself, has a sufficiently large horizon to think as Job thinks. Only the wider horizon supplied by the wisdom of the ages will do as the foundation of human thought. He now sees that position as untenable. If a single man's horizon is necessarily too confined it can only mean that the human horizon as such, including even that of the fathers, is defective. Man is a worm and has only a worms-eye view.

Nevertheless, he has not condemned Job. Why not? Is it because he too knows that he cannot prove that he is right and hence that Job is wrong?

Job sees no place that can be called a place. He says:

> I walked in gloom with no sun above. I stood up in the assembly and cried out, and so I became a brother of jackals. [30:28]

Like Job, the jackal lives on the edge of the desolate city. Perhaps Job now has no choice other than to abandon both society and his own sociality and meld into the world that knows no human eye or human tongue. For him it is a frightening thought, but the reader knows that it is into just such a world that the voice in the tempest will soon beckon him.

Then comes Elihu.

At the heart of Elihu's understanding of man is the notion that Job's request cannot be fulfilled. Man has not the stamina to face what lies beyond his own horizons, and Elihu offers himself as one who, knowing of the terrors that lie beyond the human sphere, can listen to Job's case from within the human sphere. Job cannot go beyond, nor is there any longer a need to do so.

> If there only were a messenger, an interpreter, one in a thousand, to tell man (לְאָדָם) what is right, He would have mercy on him and say: "Redeem him from descending into the muck, for I have found his ransom. Let his flesh become brighter than youth, and let him return to his springtime days." Let him but supplicate unto God and he shall be accepted and see his face with·shouts of joy, for He shall return to mortal man his sense of righteousness. Let him only stand squarely in front of mortals and say "I have sinned; I have dealt perversely with what was right, and my accounts have not been settled." Thus He shall redeem his soul from passing into the muck; and his life shall see in the light. [33:23–28]

This 'interpreter' would conquer all the frightful things that live in the world beyond the world of man. The terrible face of God would now become the home of joyous man. Job would stand where he could not stand before, and laugh where there had been only horror, because the horror would have been banished. Job, for his part, has only to confess to a sin that by its nature he does not and cannot know he has committed. Job's old hopes, that the human perspective has a legitimate place in any ultimate account of things, will still be "all heaped together in the muck," but his soul would be redeemed and pass beyond it.

For Elihu the cosmos is wholly indifferent to the fate of man and to human concerns. The pull Job felt towards a realm larger than the human realm is dangerously misdirected. In flirting with the land of the jackal, Job is allowing himself to become enmeshed in a world in which the word "justice" is a mute sound which symbolizes nothing.

> "Behold," Elihu says, "God is exalted but we cannot know. The number of His years cannot be unearthed. He draws up droplets of water, and the moisture refines itself into His mist that flows together into the nebula and trickles back down upon mankind. Who can comprehend the expanse of the clouds, the roarings under His canopy?" [36:26]

According to Elihu, the visible universe beyond man is directed by the hand of God solely towards the good of man and his concerns. Nevertheless, if Job is tempted into that world, he will be faced only by the nebula and constant roaring. To man, God's labyrinthine complex of delicately interwoven ends will seem no more than a roaring mass of anger, an incoherent vagueness. Elihu's last words are:

> The Almighty—none will find Him. He is ever multiplying in might and in right, abundant in judgment, giving neither wrack nor reason. Thus mortals hold in fear the one whom even the wise of heart have never seen. [37:23]

This is the last time that the word we have translated "mortal" will appear in the text. Etymologically, it is in fact rather close to the English word "mortal."

The root means *weak* or *sick* or, when used of a wound or a disease, it means *incurable.*

> And then the Lord answered Job out of the tempest (הַסְּעָרָה) and said: "Who is this one that makes counsel dark by words that have no meaning? But come, gird up your loins like a hero." [38:1–3a]

In contradistinction to the word *man* (אָדָם) that Elihu had used, the the voice from the tempest presses Job to be a *hero* (from the root גבר, meaning *to be strong*). The voice seems to be reminding Job of his own first words, urging him yet to fulfill the promise inherent in his birth:

> Job answered and said "May the day of my birth be lost, and with it that night in which it was said 'a hero has been conceived.'" [3:3]

and now Job is being asked to gird his loins and become that hero.

What more can be said? If Elihu were not nearly right, there would be no need for Job to gird his loins; and yet, if Elihu were simply right, perhaps there would be no point in doing so.

The voice first asks:

> I will question you, and you must let me know. Where were you when I laid the foundations of the earth? Speak up, if you know! Who fixed its measurements, if you have any understanding? Who stretched a measuring line round it, and into what were its pylons fixed? Who set the cornerstone as the morning stars sang together, and all the sons of God shouted for joy? [38:3]

But then it says:

> Who closed up the sea behind the double door when first it burst out of the womb, when I clothed it in a cloud and swaddled it in mist? [38:8–9]

This word, *to burst*, is often used of the raging seas and the monsters within them as well as of a child as it bursts forth out of its mother's womb. Here the beginning of all things is not presented as a "Let it be" or as a "creation" or a "making." This passage, mixing the waters of birth with the primordial waters of chaos, presents God more as a midwife, controlling the birth and letting it come forth in number and order. These verses, with words like "burst," "womb," "cloth," and "swaddle," seem to mark a movement from the arts to those things which are older than the arts. This movement is in fact a movement away from the masculine workman and his arts, which causes things to come to be by the application of measurement to a world outside of itself, and by the forceful fixing of pylons. In Genesis, the artisan God has within himself the "to be" of the object. He shapes and molds according to His plan; while the more feminine, swaddling, nurturing God we meet in the Book of Job allows for the emergence of the "to be" which is in the thing itself.

"Have you", the voice then asks, "yet commanded the morning, or taught the dawn to know its place, to grab hold of the corners of the earth and winnow out the wicked? All is as transformed as clay stamped by a seal, and fixed as dye in a garment.[38:12]

While the imagery of *clay* often appears in the other books of the Bible, especially in Isaiah and Jeremiah, in them, man is constantly likened to the clay itself, whereas in this speech he is like the object made of clay and bearing a seal or signet. The implication is that a man, like the pot, but unlike the clay, has his own shape, his own "to be".

In that sense, a seal marks a thing's character, and becomes a kind of guarantee that the object is what it is and will continue to be what it has become, "fixed as dye in a garment." The argument seems to be that if a white garment has been dyed red, and the dye stands firm or has been well fixed, then the garment has truly become a red garment. The red color is as much a part of the object in front of me as anything else. It would be wrong of me to think of the white garment as the truth lying behind the red garment. In the same way, the clay cannot be considered to be the truth behind the pot. The seal, or mark, or character, has transformed the clay into a pot, and a pot it now is.

The argument as a whole is intended as a reply to Elihu. The world beyond man, the realm of the jackal, as it is, considered in itself and by the human eye, may not be reducible to the nebula. The beings in it may each have their own being, and hence their own strivings and ambitions apart from any human concern; yet man may be open to them.

◆

In the next chapter Job will enter far into what we have come to call "The Land of the Jackal." There he will meet six sets of wild beasts. From each he will learn many things, but we shall only have time to meet one, the Hind.

> Do you know when it is time for the mountain goat to drop? And have you watched the hind writhing in the dance of birth? Can you number the months they fulfill? And do you know the season for them to deliver, when they couch and split open to give birth to their young, and thus to end their travail? [39:1]

The single Hebrew word חול (khol), which I have generally translated *writhe*, and here by the phrase "writhing in the dance of birth," is a very complex word. Indeed, much of our understanding of the Book of Job will center around our attempt to regain the sense of unity that lies within the complexity of this word.

As far as one can tell, it originally meant *to whirl*. It can also mean *to dance*. Sometimes it is used in a perfectly wonderful context which can be full of joy and exultation; but more often things get out of hand, so that when coming upon the word, the reader can feel a foreboding thought thickening the air. It then comes to mean *to tremble* or *to quake*, and it often means *anguish* and *pain*. But, as in our case, it can also mean *to be in labor* and hence *to give birth*. For this reason, the same word that meant *pain* and *anguish* can also mean *to prosper*.

We can now begin to understand the great admonition, "Gird your loins like a hero." Wildness and pain are present when the signet is put to the clay to make a thing of value and worth. Here there is no indication that the pain of birth is caused by a·curse or is the result of having taken a bite of the apple. Job, in visiting the dance of birth, is revisiting the day of his own birth. To venture beyond the realm of man, and to see each thing as having its own signet, means to come to terms with the unity of all these opposing meanings of חול.

> Behold, the storm of the Lord! Wrath has gone forth, a whirling (מִתְחוֹלֵל) tempest; it will burst upon the head of the wicked. [Jer. 23:19]

In this passage we may recognize some foundation for the shift we had already begun to feel in the role of the feminine. It is the whirling, dancing, pain-ridden, birthing *tempest* that speaks to Job in verse 38:1. *She* (for the word סְעָרָה, *tempest*, is a feminine noun) lets Job see that pain and joy and birth are so interrelated that they cannot be distinguished even in speech.

But we must also acknowledge another aspect of things, far removed from the intimate world of birth and dance.

> The great Leviathan counts iron as straw, and bronze as rotten wood. No son of the bow can put him to flight. Slingstones turn to stubble and clubs are rated as straw. He laughs at the sound of the javelin. [41:19–21]

This grand beast, above and beneath all malice or ambition, oblivious to all, rules all by the mere weight of his being. In him we recognize our limitations and hence see our definition.

The world beyond man is not a world in which man can play, as God can play. Man cannot toy with Leviathan as like a bird, or tie him on a string for young girls (40:29); it is only a world in which man can *learn* about play. The charm of the image of Leviathan as a bird on a string teaches us about innocent jesting; but *the world beyond ourselves* is not ours. We cannot divide it up and use it as we will. To see it at all is to see it as a thing for itself, not as a thing for us.

> Then Job answered the Lord and said, "I know that you can do all, and that no design can be withheld from you. 'Who is this one that hides counsel without knowledge?' I have spoken though I had not understood. There is a world beyond me, a world full of wonders that I had

> never known. Now listen and I will speak; I shall question you, and you will inform me. I had heard of you as ears can hear; but now my eyes have seen you. Wherefore I have both contempt and compassion for dust and ashes." [42:3]

Job the homeless is now at peace in a very large world in which no man counts for more than a hill of beans. He is also at peace in a very small world in which each man is of infinite value. He can be at peace in each world only because he is at peace in the other. He also knows that that large and woolly world has in it a kind of love and a kind of laughter which only he and his fellows can establish in the world of the small.

Job's eyes have seen; but that seeing took place in a foreign land where his hands could not act. Job of the wide world is again Job living in a nutshell, the servant of the Lord. He who has seen Leviathan will say a prayer for his friends as they bring their bulls and their rams to be sacrificed. The world of seeing has turned out to be a world devoid of all meaningful human action, and Job has returned.

His last great act is recorded:

> In all the land there were not to be found any women more beautiful than Job's daughters, and their father gave them an inheritance alongside their brothers. [42:15]

Nothing is said about husbands or dowries; the inheritance is outright and absolute. So far as I can see, that means Job has established the right of women to own and hold property. This is part of a legacy that Job has brought back from a very large world to the very small world.

Job was caught in a clashing of worlds. One pronounced "just" while the other cried "unjust." The need for human friendship and the need for clarity refused to mesh. That need for clarity, which came about when Job's world began to fall asunder, led him ultimately to the need for autonomous understanding, and hence to questions concerning those old accounts of "the first things."

Out of the whirl of the tempest came the notion of the *signets*, the understanding that things had their own seal upon them and that thery were what they were in themselves, apart from human need. This recognition led to a shift in Job's understanding of, and sensitivity to, beauty. The needs of man may better be served by being open to the excellence of things as they grow of themselves than by seeing them as being directed towards those needs. This insight, in turn, led, as we have seen, to the emergence of the nurturing and swaddling God, as distinguished from the making and constructing God. A small change in a last will and testament was the result.

The Mazkir

Maurice Ravel's Bolero begins in an idyllic mode. It establishes a rhythm that turns about itself, easily leads back into itself, with no outside force impinging. The incessant pounding never lets up. A counter-rhythm comes and goes, but the fundamental rhythm, with its required precision, never abates. The demand begins. Stronger and stronger it grows till its unrelenting precision leads on to its own chaotic destruction.

The Book of Judges also begins in an idyllic mode.

> After the death of Joshua, the people of Israel inquired of the Lord, "Who shall go up first for us against the Canaanites, to fight against them?" [Judg. 1:1]

These verses remind us of another time and another place. When Moses was about to die, he said to the Lord,

> Let the Lord, the God of the spirits of all flesh, appoint a man over the congregation who shall go out before them and come in before them, who shall lead them out and bring them in, that the congregation of the Lord may not be as sheep that have no shepherd. So the Lord said to Moses, Take Joshua the son of Nun, a man in whom is the Spirit, and lay your hand on him. [Num. 27:16]

Even that first time, the Lord seemed slow to answer, and it was Moses who took the initiative. This time the Lord says

> Judah shall go up: behold, I have given the land into his hand. [Judg. 1:2]

Strangely enough, it looks as if, again, the Lord has appointed a single human being, Judah, to *go up first.* Everything is stated in the third person singular just as before, but the man Judah had been dead for many years. The Lord was speaking of the whole tribe as if it were a single person. In this idyllic world, the two tribes, Judah and Simeon, continue to speak to each other in the singular, as if each were one, and both were brothers. The world in which these men live is wonderfully captured in speech by a phrase which occurs and reoccurs in the final chapters of the book, but by then all its charm will be gone, and we will be left with nothing but the horror. *There was no king in Israel. Each man did what was right in his own eyes.*

Now, if we compare Judah's adventures with Chapter Fifteen of the Book of Joshua, we can see that he was able to return to the battle field and accomplish what Joshua had been unable to accomplish in his day. These thoughts lead

JOSHUA 15	JUDGES 1
And he went up from there against the inhabitants of Debir. The name of Debir formerly was Kiriath-sepher. And Caleb said, "Whoever strikes Kiriath-sepher and captures it, I will give him Achsah my daughter as wife." And Othniel the son of Kenaz, the brother of Caleb, captured it. And he gave him Achsah his daughter as wife. When she came to him, she urged him to ask her father for a field. And she got off her donkey, and Caleb said to her, "What do you want?" She said to him, "Give me a blessing. Since you have given me the in land of the Negeb, give me also springs of water." And he gave her the upper springs and the lower springs. But the Jebusites, the inhabitants of Jerusalem, the people of Judah could not drive out, so the Jebusites dwell with the people of Judah at Jerusalem to this day.	And they went up from there against the inhabitants of Debir. The name of Debir formerly was Kiriath-sepher. And Caleb said, "Whoever strikes Kiriath-sepher and captures it, I will give him Achsah my daughter as wife." And Othniel the son of Kenaz, the younger brother of Caleb, captured it. And he gave him Achsah his daughter as wife. When she came to him, she urged him to ask her father for a field. And she got off her donkey, and Caleb said to her, "What do you want?" She said to him, "Give me a blessing. Since you have given me in the land of the Negeb, give me also springs of water." And Caleb gave her the upper springs and the lower springs. But the Jebusites, the inhabitants of Jerusalem, the people of Benjamin could not drive out, so the Jebusites dwell with the people of Benjamin at Jerusalem to this day.

the author back to the days of Caleb; and the rest of Chapter One of Judges is almost an exact quote from that time. One critical correction, however, has been made, which one can see by comparing Judges 1:21 and Joshua 15:36 (top of page). What that "correction" means, and how critical it was, will become clearer in later chapters.

This going back not only introduces us to Othniel, who will play an important role later in the chapter, but more importantly, it reconfigures the past so that it was Benjamin, the tribe of Saul, and not Judah, the tribe of David, that failed to conquer the city of Jerusalem. The strictly parallel nature of the relationship between the two passages over so many verses makes the switch seem the more intentional on the part of the author. As we shall see in a number of cases, The Book of Joshua is often taken to be the account of things as they chanced to have occurred, whereas The Book of Judges is intended to articulate the way things should have been, or rather the way the world fundamentally is. This reconfiguration of the past will play a larger part in our story as it unfolds.

Chapter Two begins with the voice of an angel, reminding Israel of how they were brought out of Egypt to begin a new world dedicated to the Lord, and of their failure to accomplish that end. The new world will not happen, and the old people living in the land will continue to be a thorn in their side. Verses 6 through 10 return to the death of Joshua. We have here a new beginning; but it is not the idyllic road that we had been shown in Chapter One. As opposed to that idyllic world, Verse 11 reads,

> And the children of Israel did what was bad in the eyes of the Lord, and they served the Baalim. And they left the God of their fathers. [Judg.2:11]

This statement, *did what was bad in the eyes of the Lord,* will reappear eight times during the course of the book, but, interestingly enough, the term *sin* will never appear; though, in two places, the children of Israel will accuse themselves of having sinned. The remainder of Chapter Two consists of a brief summary of the Book of Judges as a whole. God will appoint judges from time to time, to replace the idyllic life of brotherhood. Things will go smoothly enough as long as the judge is alive; but when the judge dies, the world will fall back into chaos again. Thus, we are told in advance that this four-hundred-year experiment will fail. Our only job is to see exactly how and why.[1] The first such judge was Othniel. Though he was clearly the successor to Joshua since he was the son of Joshua's brother Kenaz, the text goes out of its way to make clear that he did not automatically become judge after the death of Joshua, but only after; and for the second time, the people of Israel did what was *bad in the eyes of the Lord.* Without that, there might never have been any need for a judge.[2]

When, for the fifth time, Israel *does bad in the eyes of the Lord,* God sends the Midianites, whose rule is disastrous. Then God sent an angel to a most unlikely man, Gideon. When we first meet Gideon, he seems to be inside a building, which I take to house a wine press, beating out wheat. Most people translate:

> Gideon was beating out wheat in the winepress to hide it/him from the Midianites [Judg. 6:11]

even though there is no word or even an objective ending corresponding to the word for the assumed direct object *it/him.*

The word that the text uses for *hiding* is normally a transitive verb, and most people take it to mean that he is either hiding the wheat or hiding his father, Joash; but the use of the word in the next chapter shows that the word can also be used intransitively where it is taken to mean *to flee*:

> Every man stood in his place around the camp, and all the army ran. They cried out and fled.

This seems to imply that Gideon was in the winepress lying low, fleeing from the Midianites. A somewhat more literal translation might then be called for:

1 Judg. 2:16–23

2 Judg. 3:7–11

> Gideon was beating out wheat in the winepress to hide from the Midianites.

Next an angel appeared to Gideon and, partly in irony and partly to bolster Gideon's spirits, said to him:

> The Lord is with you, O mighty man of valor .... Go in this might of yours and save Israel from the hand of Midian; do not I send you?

There follow a number of impressive victories, in which both God and Gideon go to extreme ends to make it appear that the victories were not gained by human courage, but by God's miraculous intervention.[3] Nonetheless, the men of Israel—with a good heart, I believe—insisted upon attributing the victories to Gideon and demanded that he accept kingship. So great is the human need for human heroes! They say this to Gideon:

> Rule over us, you and your son and your grandson also, for you have saved us from the hand of Midian. [Judg. 8:22]

Gideon, then, replies to them with what seems to be the overriding concern and most fundamental teaching of the book up to this point:

> Gideon said to them: "I will not rule over you, and my son will not rule over you; the Lord will rule over you." [Judg. 8:23]

For Gideon too, the world in which *there was no king in Israel; each man did what was right in his own eyes* is an idyllic world; but this time, we are beginning to see why a world so composed cannot survive.

Gideon's position is, however, fully in accord with Samuel's most cherished thoughts, which he held on to even after he had already anointed Saul as king.[4]

The text continues:

> Jerubbaal the son of Joash went and lived in his own house. Now Gideon had seventy sons, his own offspring, for he had many wives. And his concubine who was in Shechem also bore him a son, and he called his name Abimelech. [Judg. 8:29–31]

Peace having been restored, the text goes on to describe what appears to be the well-earned rest of a great hero; but this line from this the Book of Judges must give pause to the close reader: *Gideon had many sons.* One of them, Abimelech, was the son of a concubine from the city of Shechem. The reader remembers that Shechem was a place where Simeon and Levi once killed their newly adopted brothers. It was also the place where the brothers almost killed their brother Joseph. And, finally, it is the place where the bones of Joseph lie, as if he had in fact been killed there.[5]

---

3 Judg. 7:15

4 I Sam. 12:11

5 Gen. 34:24, 37:12, Josh. 24:32

Now this Abimelech was able to persuade the men of Shechem to make him king by threatening them with the notion that the sons of Gideon would take kingship over them, and that he, Abimelech, one of their own since his mother was from Shechem, would be able to ward off the kingship of the other sons of Gideon. All this in spite of the fact that Gideon had rejected the notion of kingship, both for himself and for his descendants. "Which is better for you," Abimelech asked, "that all seventy of the sons of Jerubbaal rule over you, or that one rule over you?" The reader, hearing the answer which the men of Shechem give to Abimelech, cannot hear the words "He is our brother" without feeling the full weight of the author's sense of irony, given what brotherhood means in the city of Shechem.

Having convinced his brothers on his mother's side to follow him as king, Abimelech went to the house of his father and killed all fifty of his brothers on that side of the family. Or, rather, he killed all but one, the youngest, Jotham. Jotham then ran up to the top of Mount Gerizim and presented the most compelling argument for the general thesis of the book as it has unfolded thus far. Jotham's argument takes the form of a parable concerning the trees"

> The trees once went out to anoint a king over them, and they said to the olive tree, "Reign over us." But the olive tree said to them, "Shall I leave my abundance, by which gods and men are honored, and go hold sway over the trees?" And the trees said to the fig tree, "You come and reign over us." But the fig tree said to them, "Shall I leave my sweetness and my good fruit and go hold sway over the trees?" And the trees said to the vine, "You come and reign over us." But the vine said to them, "Shall I leave my wine that cheers God and men and go hold sway over the trees?" Then all the trees said to the bramble, "You come and reign over us." And the bramble said to the trees, "If in good faith you are anointing me king over you, then come and take refuge in my shade; but if not, let fire come out of the bramble and devour the cedars of Lebanon." [Judg. 9:7–16]

Each tree has its own particular activity which it enjoys, and with which it fulfills the needs of the world as a whole, and which gives joy to both man and God. If each tree does its own work there is no need for a king. Only the bramble, which has no proper activity of its own, is willing to be king; but he can only rule as a tyrant, since he has no virtue of his own.

This parable seems to me to be yet another way of praising a world in which *there was no king in Israel; each man did what was right in his own eyes.* Yet the contrast between that ideal and the world actually facing us is frightening. After Abimelech the Israelites had a period of twenty-five years of peace under two judges; but then they again *did what was bad in the eyes of the Lord.* This time they were attacked by the Ammonites who invaded the city of Gilead. Its founder, Gilead, had once had a son, Jephthah, known as a mighty warrior, the son of a prostitute. Some years later, when the sons of Gilead's wife grew to adulthood, they drove out Jephthah; but after the Ammonites attacked, they pleaded with Jephthah to return as their leader.

In his zeal, Jephthah felt that a deep sense of cosmic order demanded that he sacrifice some dear thing for the Ammonites to be given into his hand. This strange, even grotesque, sense of balance led him to vow to sacrifice as a burnt offering whatever came out of his door to meet him when he would return in peace from the Ammonite war. We are both charmed and horrified when his only daughter greets him with laughter and dancing with a tambourine. Both father and daughter, so full of life and springtime, are horrified and saddened by what is about to happen; yet both feel a deep sense of duty to do what was right in their own eyes.

Then we go right on to the next story. As is so often the case, the story of Samson begins by the announcement that a barren woman shall conceive and bear a son.[6] This time, however, things are different, and they do not proceed quite as smoothly as they had for others. An angel appeared to the woman, but when she told her husband Manoah that a "man of God" had appeared to her, he was somewhat unclear as to how to proceed, and asked the Lord to let the messenger return; but when the angel did return he chose to appear again to the woman, and she was forced to call her husband to speak with the him. From the conversation, one can tell that the angel was somewhat miffed at having to repeat the promise and its attendant conditions. But he saw that Manoah was a kind and extremely generous man, even if his generosity was a bit awkward and somewhat misplaced; and the angel departed by putting on a great show for him.[7]

When Samson grew to manhood, he

> went down to Timnah, and at Timnah he saw one of the daughters of the Philistines. Then he came up and told his father and mother. "I saw one of the daughters of the Philistines at Timnah. Now get her for me as my wife." [Judg. 14:1]

This is not the first time that a man of our story met a woman from the small town of Timnah and took her to be a prostitute. Last time it was Judah; this time it is a man from Dan.[8] Memories of that first encounter which led to the birth of Perez and so to the house of David, give rise to great hopes and expectations in the reader's mind; but this time those hopes will not be fulfilled.

---

6 Judg. 13:1–7; Gen. 11:30, 17:16, 25:21, 29:31, 30:22; I Sam. 1:19

7 Judg. 13:13–23

8 Gen. 38:1–30, Ruth 4:18

Samson's parents, like Abraham and Isaac before them, are concerned about their son's choice of a wife.[9] Samson's answer to his father is no more than that haunting phrase, *she is right in my eyes*, which play such an ambiguous role in this book.

Some days later, Samson went down to Timnah to prepare a feast for his new friends, the Philistine comrades. On the way he found that a swarm of bees had made their home in the carcass of a lion that he had ripped apart with his own bare hands a few days before; and now the bees had filled it with honey. He took a bit of it and gave it to his parents, but didn't tell them where it had come from. At the feast, Samson bet his friends that he could ask them a riddle which they could not answer. The bet was for thirty linen garments and thirty changes of clothing; they answered:

> Put your riddle, that we may hear it.

The riddle was

> Out of the eater came something to eat.
> Out of the strong came something sweet.

There was, of course, no way they could guess the answer, so they turned to Samson's wife and threatened her life if she did not entice her husband to give them the answer to the riddle. Seven days she wept and pleaded, and finally he said:

> What is sweeter than honey?
> What is stronger than a lion?

Samson's wife reported this to her people, who thereby answered the riddle. Samson, for his part, became enraged:

> And the Spirit of the Lord rushed upon him, and he went down to Ashkelon and struck down thirty men of the town, took their spoil and gave the garments to those who had told the riddle. [Judg. 14:1–5]

From these actions we can begin to see how the Spirit of the Lord articulates itself in Samson; and, especially in the light of what is to come in the next chapter, we can see with what a strange and contorted sense of justice Samson felt bound to pay his debts. Some days later, Samson stormed back to his father-in-law's house to retrieve his wife; but in the meantime his father-in-law, on the assumption that Samson utterly hated his wife for what she had done, had given her as wife to Samson's best man.

> And Samson said to them, "This time I shall be innocent in regard to the Philistines, when I do them wrong." So Samson went and caught 300 foxes and took torches. And he turned them tail to tail and put a torch between each pair of tails. And when he had set fire to the torches, he

9 Gen. 28:1ff

> let the foxes go into the standing grain of the Philistines and set fire to the stacked grain and the standing grain, as well as the olive orchards. [Judg. 15:1–5]

Apparently Samson's father-in-law acted as he did out of a certain respect for Samson. We are not told how Samson interpreted that action, but note that he finds it necessary to state the grounds of his retaliation: *This time I shall be innocent in regard to the Philistines, when I do them wrong.* It shows a strange awareness of the need for justice, regardless of how twisted that sense of justice might be.

From the next verses one can see how complicated the situation had become.[10] Samson felt it necessary to undo the wrong his comrades had done him by attacking the Philistines as a whole. The Philistines then attacked the camp of Judah, who had been totally silent since those idyllic days of the beginning. But the men of Judah react very differently from Samson. They turn to speech rather than to violence, asking the Philistines, "Why have you come up against us?" When the Philistines answer, "We have come to find Samson, to do to him what he has done to us," the men of Judah went down to Samson and demanded: "Do you not know that the Philistines are rulers over us? What then is this that you have done to us?"

When they heard Samson answer, "As they did to me, so have I done to them," Judah saw that there was not much difference between Samson and the Philistines. Then regretfully, or so it seems to me, he bound Samson and planned to hand him over to the Philistines. However, in a way that so much reminds one of the way in which Judah once saved Joseph from being killed, he promised Samson that he would not kill him. Instead, he bound him to a rock and left him there. But when the Philistines came to the rock to pick Samson up,

> The Spirit of the Lord rushed upon him, and the ropes that were on his arms became as flax that has caught fire, and his bonds melted off his hands.

When the Philistines actually gathered their force around Judah, they made a raid on the city called Lehi. They were apparently able to hold it for quite some time, since it was not till many years later, when David was an old man and about to die, that he summoned up his forces for one last battle against the Philistines. In that final battle, he at last struck down the Philistines and reconquered the city of Lehi.[11]

---

10 Judg. 15:6ff

11 II Sam. 23:11

As soon as his hands were free and Judah's battle over, Samson picked up the jawbone of an ass and struck a thousand men. Then he said:

> With the jawbone of an ass,
> A heap and a double heap;
> With the jawbone of an ass
> Have I struck down a thousand men.

To make this intelligible, it should be noted that the word for *jawbone* is a word-play on the name of the city Lehi, while the word for *heap* sounds like the word for *ass*. But despite all that bravado, it should be remembered that it was Judah that was struck at Lehi, and that it was not Samson but David that finally conquered the city. Thus we have seen that Samson's affair at Timnah, in this book, led nowhere; and at Lehi, Judah was wiser in speech than Samson, and David was better in battle.

Aside from one moment in Chapter Twenty, when God asked Judah to again take leadership, Judah plays no role in all this tumult of the Book of Judges; and, in fact, the tribe of Judah does little of note until David comes on the scene, paving the way for the new kingship.

From all this, it is clear that the special character of both Judah and Benjamin is an integral part of the foundation of the Book of Judges. And so we are forced to look at the phrase *because there was no king in Israel*, not as a sprinkling of pro-kingship material added later, but as a statement concerning the fundamental intent of the author.

Next we come to the story of Micah and his mother, a woman of strange innocence and misplaced goodwill.

> There was a man of the hill country of Ephraim, whose name was Micah. And he said to his mother, "The 1100 pieces of silver that were taken from you, about which you uttered a curse, and also spoke it in my ears—behold, the silver is with me; I took it." And his mother said, "Blessed be my son by the Lord."

Micah returned the money; and when she said, "I dedicate the silver to the Lord from my hand for my son, to make a carved image or a cast metal image," it seems almost as if she had never heard, or did not understand, or perhaps was not even capable of believing that God had really meant it when He said:

> Cursed be the man who makes a carved image or a cast metal image, an abomination to the Lord, a thing made by the hands of a craftsman, and sets it up in secret. [Deut. 27:15]

Micah's mother gave money to the silversmith:

> And the man Micah had a shrine, and he made an ephod and household gods, and ordained one of his sons, who became his priest,

and we remember Gideon. The chapter ends with that taunting phrase that will continue to ring throughout the rest of the book,

> In those days there was no king in Israel. Everyone did what was right in his own eyes. [Judg. 17:1–60]

Here we can see how carefully the author chose his words, and in what sense Micah's mother was indeed faithfully trying to do what was *right in her own eyes.*

The next part of our story again begins with those fateful words: "In those days there was no king in Israel." The tribe of Dan had not yet found an adequate place for themselves, so they decided to send five men up to the north country to scout for more land.

> Then the five men departed and came to Laish and saw the people who were in it living in security, after the manner of the Sidonians, quiet and secure; for there was no ruler humiliating them for anything in the land, and they were far from the Sidonians and had no dealings with anyone. [Judg. 18:7]

The people of Laish seem to be living that never-to-be-lived idyllic life which we saw, though only momentarily, at the beginning of the book. There is no king in Laish, yet they live securely; and for them, being ruled would be shameful and humiliating. They live quietly. No one knows they are there, and they have need for nothing outside of themselves. This would seem to be an exemplary instance of the highest life, hiding in the phrase *there was no king in Israel; everyone did what was right in his own eyes.*

The five spies, returning to camp, urged their brothers to arise and attack. They marched off to Laish with six hundred men of the tribe of Dan, armed with weapons of war. The utopian city of Laish easily fell to the first outside force that happened along—a victim of the tribe of Dan. It is, however important to note that the idyllic nature of these people is a pure conceit of the Book of Judges. There is nothing even remotely idyllic to be found in the parallel passage from the Book of Joshua concerning Laish, or Leshem as it is there called. It is a town of no particular importance and is never mentioned again.

> When the territory of the people of Dan was lost to them, the people of Dan went up and fought against Leshem, and after capturing it and striking it with the sword they took possession of it and settled in it, calling Leshem Dan, after the name of their father Dan. [Josh. 19:47]

It is by means of such parallel passages as this that the biblical author is able to articulate what is most revealing in the general fabric of the world and distinguish his account from the chance course of time. We have seen this relationship between the two books before, when it became Benjamin and not Judah who failed to capture Jerusalem. The text continues:

> Then they rebuilt the city and lived in it. And they named the city Dan, after the name of their father Dan, who was born to Israel; but the name of the city was Laish at the first.

How innocent that renaming seemed back in the Book of Joshua; how chilling now. And all that remains is this gentle reminder that there was once a wondrous city named Laish. The tale of Micah ends with the words:

> And the people of Dan set up the carved image for themselves, and Jonathan the son of Gershom, son of Moses, and his sons were priests to the tribe of the Danites until the day of the exile of the land. So they set up Micah's carved image that he made, as long as the house of God was at Shiloh. [Judg. 18:30–31]

What is meant by the phrase *until the day of the exile (or captivity) of the land?* In fact, the New Revised Standard Version translates: *...until the time the land went into captivity.*

Some people see in the words "until the day of the exile of the land" a reference to the dying words of the wife of Phineas: *The glory has been exiled from Israel* (I Sam. 4:22). But the words *the day of the exile of the land* say nothing about *glory*. Rather, our passage seems to be a reference to a verse from II Kings:

> So Israel was exiled from its soil to Assyria until this day. [II Kings 17:23]

What, then does that mean for our reading from the Book of Judges? The book would then be exactly what it appears to be to the eyes of the naive, inexperienced first-time reader. That is, the book would appear merely as one in a series of books, each beginning by clearly referring to the one before it, and ending by introducing the next: Deuteronomy ends with the appointment of Joshua, the Book of Joshua ends with Joshua's death, the Book of Judges clearly begins with the death of Joshua and ends with the need for a king.

This would imply that while the Book of Judges looks forward to the establishment of the kingship, the reference to *the day of the exile of the land* implies that its author is already aware of the fact that kingship would ultimately fail. If we look at the series of books as a whole, then, we see that the People of Israel began in slavery in Egypt, and that their story ends again in slavery (this time in Babylon) with the hope of a new beginning.

What name shall we give to the man living in Babylon who put all these things together? Today we call him the Redactor; but in Biblical Hebrew he would have been called the Mazkir, from the root *zkr, to remember*. The word Mazkir is in the causative form. To give the reader some idea of what that means, the causative form of the word *go* means *to cause to go*, that is, *to bring*. However, the causative form can have a secondary significance. The causative form of the word *to be just* means *to declare a thing just*. Then the question arises, shall we call the Mazkir *the Reminder*, or something like *He-who-fashions-our-memories*? Perhaps we ought simply call him the Mazkir.

Is it possible that the same author wrote both these passages?—

> But the Jebusites, the inhabitants of Jerusalem, the people of Judah could not drive out, so the Jebusites dwell with the people of Judah at Jerusalem to this day. [Josh. 15:63]

and

> But the Jebusites, the inhabitants of Jerusalem, the people of Benjamin could not drive out, so the Jebusites dwell with the people of Benjamin at Jerusalem to this day. [Judg. 1:21]

In other words, does the author wish us to see Judges 1:21 as a reconstruction of history? Did he wish to articulate for the reader the proper root of kingship and clearly distinguish it from the accidental or merely factual account? This is not the first time that such a thing has occurred. In the book of Genesis, we are told that Abram came from Ur of the Chaldees (Gen. 11:31). Then the years rolled by—four hundred years from Abram to Moses, four hundred years under the Judges, and four hundred years under the kings—and not a word to be heard of the Chaldeans. Then, in the final chapter of the Book of Kings, the Babylonians, swooping down, conquering and destroying the whole of what had so carefully been planned from the beginning, brought along with them their Chaldean allies; and it is they, the Chaldeans, who physically put their hands on the king and take him into captivity, destroying the holy places (II Kings 25:1–7).

For the Mazkir, there are always those ancient roots which stretch back beyond the Fathers and come back to haunt. Even though he wrote the words,

> Now the Lord said to Abram, "Go from your country and your kindred and your father's house to the land that I will show you" [Gen. 12:1],

he seems to have known at the time that ultimately our pagan past would return one day and demand to be dealt with.

Clearly, The Mazkir must have had before him many old manuscripts in which he did his research in weaving his tale; but the question arises as to whether any one of them ever claimed that Abram came from Ur of the Chaldees prior to the time of the destruction when the Chaldeans arrived in force—or was that too a conceit of The Mazkir? In order to investigate this question, it is important to realize that the Hebrew word for Chaldeans is not Kaldim, but rather Kasdim; they are the descendants of Kesed. Very late in his life, just after the binding of Isaac, and just before the death of Sarah, the story of Abraham was interrupted for no apparent reason by the passage:

> It was told to Abraham, "Behold, Milcah also has borne children to your brother Nahor: Uz his firstborn, Buz his brother, Kemuel the father of Aram, [and] Kesed. [Gen. 22:20]

This might lead one to think that the Mazkir wanted us to know that Abram could not have come from Ur of the Chaldees, because he himself has let us know that their father Kesed had not yet been born. Why the Mazkir may have chosen to do that becomes clear by comparing the account of Leshem in the Book of Joshua with the account of Laish in the Book of Judges. Had the author not given us an historical account in the Book of Joshua, we might have mistaken the account of Laish for a mere fact, one that just happened to occur by chance one day, rather than seeing it as touching the very fabric of the world. Every myth is a story, and every story happens on a day; but it must be clearly seen as never happening on any specific day, precisely because it happens every day. Without that ambiguity, the heart of the world would flit by as a passing moment in time.

Homer forged memory for the Greeks by crafting for them a battle in which the Western world, the world of the Greeks, pulled itself together through a victory over the Eastern world at Troy—a battle fought not for wealth, but for beauty. From childhood to adulthood, Greeks continually learned who they were from what they had been, as found in Homer's gift. The Mazkir has a very different kind of story to tell, and a very different kind of memory to forge. His story is the story of a people who began as slaves in Egypt, in the West, and finished as slaves in Babylon in the East. Along the way, many mistakes were made, and were recorded and refashioned. The first time, God was there to bring them out of the land of Egypt into the land of Israel. The second time they will have a *book*, along with the wisdom that comes only from defeat; and they will try again.

The final story of the book begins with those fateful words: "In those days there was no king in Israel":

> A certain Levite was sojourning in the remote parts of the hill country of Ephraim, who took to himself a concubine from Bethlehem in Judah. And his concubine was unfaithful to him, and she went away from him to her father's house at Bethlehem in Judah, and was there some four months. [Judg. 19:1–2]

This story will not repeat the kind of insanity that we have seen before and will see again, as the opening words might portend, yet the actions of an unfaithful woman add to the reader's foreboding.

> Then her husband arose and went after her, to speak kindly to her and bring her back. He had with him his servant and a couple of donkeys. And she brought him into her father's house. And when the girl's father saw him, he came with joy to meet him. [Judg. 19:3]

The husband's good will and the warm welcome tend to make the reader forget those dire words with which our story began. The girl's father is a splendid host, and the time they spend together eating and drinking is almost idyllic; yet those opening words will not go away.[12]

Four days later they finally began their return, but it was a long, hard journey; as the sun began to set, they decided to spend the night on the road. The servant wanted to spend the night in Jerusalem, the future home of king David; but Jerusalem was still in the hands of the Jebusites, and the husband, not wanting to spend the night in a city held by foreigners, chose to spend the night in Gibeah, the future home of King Saul. This despite the fact that the one Jebusite that we ever actually meet turns out to be quite a decent human being, named Araunah.[13]

But perhaps it was an error to speak of *the future home of King Saul*. Perhaps it would capture thought more accurately to speak of *the home of King Saul* simply. The notion of the future, and hence of time itself, might not have been for the Mazkir what it is for us. While, strictly speaking, there is no Biblical word for *future*, the Biblical word עָתֹד, meaning *to be ready* or *to be prepared,* soon developed a noun form, עָתִיד, which pretty much means *preparedness.* The phrase, עָתִיד לָבֹא (preparedness for what is to come) is used in the Mishnah in regard to the coming of the messiah; thus *the future*, as here understood, is *a present readiness for what is to come.*

In that sense, Gibeah was essentially always the home of Saul, just as Jerusalem was always the home of David even when it was in the hands of the Jebusites. In line with that, our own word *future* ultimately comes from a Greek word φύω, *to grow*. For the Mazkir, places bear the seed of what is to come just as much as they bear the memory of what is past.

The husband then had a second chance. He could choose between spending the night at Gibeah, or spending the night in Ramah, the future home of Samuel the Seer; and again they end up going to the town of Gibeah.[14]

As the narrative continues, the three of them—the Levite from Ephraim, the woman, and his servant—enter the town and sit in the square. No one stops by, no one greets them, and no one invites them home until an old man

12 Judg. 19:4–9

13 I Sam. 10:26, II Sam. 24:18

14 Judg. 19:13, I Sam. 7:17

from the Levite's home tribe, Ephraim, takes them in, treats them well, and gives them food and drink. The contrast between the merriment of the house and the coldness of the street is deafening.

> As they were making their hearts merry, behold, the men of the city, worthless fellows, surrounded the house, beating on the door. And they said to the old man, the lord of the house, "Bring out the man who came into your house, that we may know him." And the man, the lord of the house, went out to them and said to them, "No, my brothers, do not act so wickedly; since this man has come into my house, do not do this vile thing. Behold, here are my virgin daughter and his concubine. Let me bring them out now. Violate them and do with them what is good in your own eyes; but against this man do not do this outrageous thing." But the men would not listen to him. So the man seized his concubine and made her go out to them. And they knew her and abused her all night until the morning. And as the dawn began to break, they let her go. And as morning appeared, the woman came and fell down at the door of the man's house where her husband was, until it was light. [Judg. 19:22–26]

Note that the old man says to the men of Gibeah, "Do with them what is good in your own eyes. He does not say, "Do with them *what is right in your own eyes*," since it is unlikely that even the men of Gibeah think that what they are doing is right, or that they are doing what they perceive to be their duty. These are the considerations that distinguish this case from that of Micah's mother or from Jepthah. In this use of the expression *what is good* we have some slight confirmation of what has been said so far concerning the phrase *what is right*; though all interpretation is at best a guess.

In the morning when he opened the door, the husband seemed strangely unaware of what was before him.[15] Whether that was because he is uncaring or inattentive, or because he was totally dazed by what he saw, we do not know; but when he returned home his actions were swift and brutal:

> And when he entered his house he took a knife and, taking hold of his concubine, he divided her, limb by limb, into twelve pieces, and sent her throughout all the territory of Israel. [Judg. 19:29]

Something very similar happened in the days of King Saul—but perhaps we should recount that story from the beginning. Young Saul, before he became king, was a decent, thoughtful son, careful of others; he cared for a donkey that went astray, and he cared for a father whose son went astray looking for a donkey.[16] One might even be tempted to call that young man "idyllic," and

---

15 Judg. 19:27–28

16 I Sam. 9:1–5

certainly very different from that pesky little kid David, who thought he could kill a giant.[17]

As night came on, a servant suggested to Saul that they go off to look for the Seer to help them find their way; and in Saul's response, you can still see the sweetness in his nature.[18] But even before Saul arrived, Samuel knew that there would be a king from Benjamin; and when he hinted that Saul would become king, Saul's answer still expresses that same mild humility.[19]

But then, as king, Saul discovered that he was to hunt down Amalek, totally destroy it, and kill their king, Agag. He did defeat the Amalekites and capture Agag, but could not bring himself to kill him. For that disobedience, he lost all he ever had: a son, a daughter, a kingdom and a life. I do not pretend to know what it was about that meeting with Amalek that caused a pleasant young man to become a brooding and vengeful, and ultimately half crazed, old man; but I believe I can point out some passages that may be helpful for facing this question.

First, it must be understood that Amalek, grandson of Esau, was the Israelites' closest living relative: closer than Ishmael, closer than the Ammonites or the Moabites, closer than the sons of Lot and the father of Ruth, and so closer than the father of David. The Amalekites were closer than the Midianites, those children of Abraham to whom Moses ran and from among whom he found a wife. They were closer than the Kenites.[20]

There had always been talk that the blessing given to Abraham would one day spread out to the entire earth. If Balaam is to be believed, it appears that Amalek was to have been the first to begin the expansion of that blessing.[21]

According to Joshua's account, it would seem that Esau was given Mount Seir, east of Egypt, so that the sons of Jacob might gain a direct path into the land of Canaan. But this never occurred, because Amalek attacked from the rear and forced Israel to fight its first war.

> At that time the Lord said to Moses, Write this as a memorial in a book and recite it in the ears of Joshua, that I will utterly blot out the memory of Amalek from under heaven. [Exod. 17:8]

Utterly blot out: this was the word used at the time of the flood; it means to *efface, erase, cancel, or obliterate*. After the flood, there was no carnage, there

---

17 I Sam. 17:32

18 I Sam. 9:7

19 I Sam. 9:15–21

20 Gen. 36:12

21 Num. 24:20

were no bodies floating by; nothing remained. The names Adam, Eve, Eden, Cain or Abel—none of them would ever again appear in the text. The world was as though that other world, the one before the flood, had never existed. This is what Saul will be commanded to do: establish a world in which Amalek never was; but this is something Saul will be incapable of understanding.

Had the Amalekite attack not taken place, it is likely that Edom would not have felt fear, and Israel might still have enjoyed a clear, free passage into the promised land.[22] Had Edom not refused to give Israel passage through his territory, none of the wars in Jepthah's account would ever have taken place. No one would have died, and there would have been peace in Jepthah's own day.

The command to Saul was clear and distinct, and fully in accord with God's command to blot out the memory of Amalek from under the heavens. These are the things that Saul, in his idyllic beginnings, failed to understand.

Before going off to war, Saul, on his own, went to the Kenites to warn them of the impending attack. To that extent, we still see that same thoughtful Saul that we have known from the beginning.[23] In the battle, Saul destroyed Amalek, but he could see no point in destroying king Agag, or the best of the cattle.

> Saul and the people took pity upon Agag and the best of the sheep and of the oxen and of the fattened calves and the lambs, and all that was good, and felt no desire to utterly destroy them. But all that was despised and worthless they devoted to destruction. [I Sam. 15:9]

For that, Saul's kingship was lost.

When confronted by Samuel with the bleating sheep, Saul seems totally confused, genuinely believing that he had fully complied with the charge laid upon him as king. From his answer, it is clear that Samuel believed Saul's claim that he intended to sacrifice the spoils; but Samuel also knew that Saul could not appreciate the distinction between *sacrifice* and *utter destruction* because he did not ultimately grasp the full implications of Amalek's actions.[24]

The formerly pesky kid David is quietly appointed king, and Saul sits and broods. Saul's temper can only be quieted by music from the very one about whom he is brooding. The knowledge of what a king must do will slowly drive the idyllic man mad, but the pesky kid will make it through.[25]

When David and Saul returned from striking down the Philistine, everyone came out from all the cities of Israel, singing and dancing, to meet them with

---

22 Num. 20:14

23 I Sam. 15:6

24 I Sam. 15:10–21

25 I Sam. 16:14–23

tambourines, with songs of joy, and with musical instruments. Then there was the women's song—a song possessing a beat in a language that rarely knows rhythm, a constant beat that drove Saul insane. Even David's song could not undo the song of the women; and we remember that idyllic child whom kingship has made mad.

> And the women sang to one another as they danced, "Saul struck down his thousands, and David his ten thousands." [I Sam. 18:7]

Near the end of his life, Saul met the Philistines one more time. The war was going badly, Samuel was dead, and Saul was frightened. At that point, the text simply says:

> And Saul had put the witches and the necromancers out of the land. [I Sam. 28:3]

The verse is so sparse at this point that it is difficult to know why Saul did so. There is, of course, a law to that effect (Deut. 18:10), but it is hard to believe that that was uppermost in Saul's mind. There seemed to be little to fear from the witches at that time; but perhaps Saul was afraid of where his own fears might lead him. He sought answers from the Lord in dreams, from the *urim*, and from the prophets; but all were silent. Saul then disguised himself and went by night to see the Witch of Endor. She raised Samuel up from the dead; but Samuel only turned away from Saul in death as he had in life. When Samuel turned away, and the woman saw that this man, the one who would have killed her, had gone insane, she treated the king as a mother would a child. But the war went on.[26] That was Saul's last battle. The Philistines stripped him and cut off his head, although he was already dead.

By contrast, a man came to David after the battle, claiming that at Saul's request he had assisted the king in his attempt to commit suicide. David could have praised the man for his service to the king; but once David learned that the man was an Amalekite, the man was doomed.

> And David said to the young man, "Where do you come from?" And he answered, "I am the son of a sojourner, an Amalekite." David said to him, "How is it you were not afraid to put out your hand to destroy the Lord"s anointed?" Then David called one of the young men and said, "Go, fall upon him." And he struck him down so that he died. [II Sam. 1:13–15]

There is more to the story of King Saul, but now we must return to the Levite and his concubine. When word went out from the Levite, four hundred thousand footmen that drew sword assembled as one man to the Lord at Mizpah, and asked "Tell us, how did this evil happen?" The Levite's account is fair and

26 I Sam. 28:8–25

straight forward except for one seemingly minor change.[27] The reader can well understand that, because of his fear and anger, it was deeply embedded within the memory of the Levite that it was the *lords of Gibeah* that rose against him. However, we know that in fact it was only some *worthless fellows*. To the Levite that difference may have been insignificant; nevertheless, if memories had been otherwise, this very bloody war might never have taken place.

But all the men of Israel were already gathered against the city of Gibeah, united as one man—a fragment of the idyllic, and we shall see how dangerous a fragment of the ideal can be.

> Then the tribes of Israel sent men throughout all the tribe of Benjamin, saying, "What evil is this that has taken place among you? Now therefore give up the worthless men in Gibeah, that we may put them to death and purge evil from Israel." [Judg. 20:12]

For better or for worse, the tribes of Israel are now united; moreover they have ceased to speak of the *lords of Gibeah*, but *the worthless men in Gibeah*. Have they somehow discovered that it was only a few worthless ruffians who are to blame? Or do they count the whole of Gibeah as just a bunch of worthless men?

And what is the duty of the Benjaminites? Should they side with their brother Benjaminites, or to their other brothers, the children of Israel? How would they know? Apparently it was right in the eyes of the Benjaminites to join their closer brothers, the Benjaminites of Gibeah rather than their more distant brothers, the people of Israel.[28]

As they did in the beginning of the book, the children of Israel ask God, "Who shall go first?" And the Lord again says, "Judah shall go first."[29] Now when Judah was first appointed leader, back in idyllic times, he performed ten separate acts within the first eighteen verses:

Judah shall go up...	Judges 1:2
Judah said...	1:3
Judah went up...	1:4
[They] found...	1:5
Judah fought...	1:8
Judah went down to fight...	1:9
Judah went...	1:10
Judah went...	1:11
They defeated...	1:17
Judah captured...	1:18

27 Judg. 20:4–7

28 Judg. 20:14

29 [60]Ju 20:18

But this time, in Chapter 20, not a single act is reported of him; we actually have no reason to believe that he even showed up for the fiasco that lies ahead.

Israel was still unsure, and they asked again: "Shall we again draw near to fight against our brothers, the people of Benjamin?" And the Lord said, "Go up against them."[30]

The Lord has sent Israel off to a war in which it will lose twenty-two thousand men on the first day and eighteen thousand on the second day. On the third day they will succeed in killing twenty-five thousand of their brothers from Benjamin, an action they will soon regret. Even a reader who sees God as the most insightful character in one of the world's most insightful books might wonder why God should do such a thing. To such a reader, I can only suggest that perhaps the Lord wanted Israel to see the full implication of the direction in which he was headed, even if his natural bent was not to go the whole way.

After the day of the second battle, the people of Israel, with tears in their eyes, inquired of the Lord:

> For the ark of the covenant of God was there in those days, and Phinehas the son of Eleazar, son of Aaron, ministered before it, saying, "Shall we go out once more to battle against our brothers, the people of Benjamin, or shall we cease?" And the Lord said, "Go up, for tomorrow I will give them into your hand." [Judg. 20:27–28]

By giving the tale of the Levite from Ephraim as the final story of the book, the Mazkir establishes it as the true culmination of the book as a whole. The charm and good-natured character of the concubine's father set a tone of idyllic conviviality; but its unstable nature was revealed in the frightful reaction good men had to an atrocious act. In this story, we shall see to what extent a resolution to the problems raised can be found within the book itself, and to what extent resolution can only be found in a later book. The final and defining character of these events for the Mazkir is independent of the fact that they happened to have occurred some several hundred years earlier, when the sons of Israel had first crossed over into the promised land, in the days of Phineas, the son of Eliezar. Josephus, the historian, however, restores them to what he believes to be their proper place at the beginning of the book.

The Mazkir, whoever he was, that man sitting and writing in Babylon, the one people today call the Redactor, must have had before him many accounts both of and from the past. Some were surely written down and lying on a table in his library; but were there others that he collected from tales handed down by word of mouth? It's hard to know. According to the Mazkir, this last story of madness and of life without a king took place in the city of Gibeah,

30 Judg. 20:23

which happens to be the home of that first king, Saul, who himself went mad. Unfortunately we do not have access to that library; but the notion that both of these events took place in the same city by mere chance is so curious, that one is forced to wonder: why do things like madness seem to take root in a place and persist there? Might it actually be rooted in the world the Mazkir has opened up for us, not just a theme he happened to find in the older accounts?

After the last battle was over, the men of Israel realized that

> They had sworn at Mizpah, "No one of us shall give his daughter in marriage to Benjamin." And the people came to Bethel and sat there till evening before God, and they lifted up their voices and wept bitterly." And they said, "Lord, God of Israel, why has this happened in Israel, that today there should be one tribe lacking in Israel?" [Judg. 21:1–3]

> And the people of Israel said, "Which of all the tribes of Israel did not come up in the assembly to the Lord?" For they had taken a great oath concerning him who did not come up to the Lord at Mizpah, saying, "He shall surely be put to death." And the people of Israel had compassion for Benjamin their brother and said, "One tribe is cut off from Israel this day. What shall we do for wives for those who are left, since we have sworn by the Lord that we will not give them any of our daughters for wives?" And they said, "Behold, no one came to the camp from Jabesh-gilead, to the assembly." For when the people were mustered, behold, not one of the inhabitants of Jabesh-gilead was there. So the congregation sent 12,000 of their bravest men there and commanded them, "Go and strike the inhabitants of Jabesh-gilead with the edge of the sword; also the women and the little ones. This is what you shall do: every male and every woman that has lain with a male you shall devote to destruction." And they found among the inhabitants of Jabesh-gilead 400 young virgins who had not known a man by lying with him, and they brought them to the camp at Shiloh, which is in the land of Canaan. [Judg. 21:4–12]

In their zeal to maintain their oath, and their great compassion for their brother Benjamin, the children of Israel are oblivious the fact that they are about to destroy a city for not having participated in an act which they themselves now regret having committed.

Soon after Samuel anointed the idyllic young Saul as king, word came that the Ammonites had besieged this town of Jabesh-gilead, and threatened to gouge out the right eyes of all the inhabitants. That happened before Saul had even heard of Amalek. When the news arrived, he was in the field plowing with his oxen like any farmer, even though he was king.[31] Saul took the field that day, and the city was saved. But years rolled by, God had forsaken Saul and turned to David; Samuel had abandoned Saul, and Saul was left to go mad.

---

31 I Sam. 11:5

The ghost of Samuel turned away from him, and the Philistines left his head hanging from a tree. But the men of Jabesh-gilead did not forget what Saul had done for them.

> They arose and went all night and took the body of Saul, and they came to Jabesh and burned it there.... And they took the bones and buried them under the tamarisk tree in Jabesh and fasted seven days. [I Sam. 31:11–13]

These valiant men were the ones who brought the house of Judah and the house of Benjamin back together again.[32]

This is a picture of the town that Israel would have destroyed in order to fulfill what it considered to be its sacred duty: both to live up to its oath and to save its brother Benjamin from destruction. All this for a sense of compassion and duty towards an oath that should never have been taken!

> Then the elders of the congregation said, "What shall we do for wives for those who are left, since the women are destroyed out of Benjamin?" And they said, "There must be an inheritance for the survivors of Benjamin, that a tribe not be blotted out from Israel. Yet we cannot give them wives from our daughters." For the people of Israel had sworn, "Cursed be he who gives a wife to Benjamin." [Judg. 21:16–18]

The elders gave to the surviving Benjamites those they had saved alive of the women of Jabesh-gilead; but there were not enough for them. One can be both impressed and horrified by the seriousness with which the elders faced their dilemma.

> So they said, "Behold, there is the yearly feast of the Lord at Shiloh, which is north of Bethel, on the east of the highway that goes up from Bethel to Shechem, and south of Lebonah." And they commanded the people of Benjamin, saying, "Go and lie in ambush in the vineyards and watch. If the daughters of Shiloh come out to dance in the dances, then come out of the vineyards and snatch each man his wife from the daughters of Shiloh, and go to the land of Benjamin. And when their fathers or their brothers come to complain to us, we will say to them, 'Grant them graciously to us, because we did not take for each man of them his wife in battle, neither did you give them to them, else you would now be guilty.'" [Judg. 21:19–21]

The wise elders of the congregation were able to find a solution to their conundrum within the parameters of their world, without ever having to question those parameters.

> And the people of Benjamin did so and took their wives, according to their number, from the dancers whom they carried off. Then they went

32 II Sam 2:4, 21:12

> and returned to their inheritance and rebuilt the towns and lived in them. And the people of Israel departed from there at that time, every man to his tribe and family, and they went out from there every man to his inheritance. [Judg. 21:23–24]

After the death of 65,130 men, when *men of the city, worthless fellows,* became first *the leaders of Gibeah,* then became *the worthless men in Gibeah,* all grew quiet, quiet as evening on the afternoon of a faun. All was forgotten, and each man went home to his tribe and his family to lead the idyllic life—because, in the final words of the book,

> In those days there was no king in Israel. Everyone did what was right in his own eyes. [Judg. 21–25]

Leo Strauss' Reading of the Bible
and its Implications[1]

As one begins to look more closely at what Mr. Strauss wrote and said about the Bible, one begins to feel a greater and greater need to consider what I must whimsically call Strauss's good-natured theory of history. I believe that what I mean by *whimsical* is clear; I hope that what I mean by *good-natured* will become less opaque during the course of this talk.

Strauss begins with the notion of man as a rational animal. By this, he meant nothing more than that by day men work together, and in the evening they talk. Of what do they speak? Of the way to plant corn, the way to go out on the·hunt; the way to bake bread, the way to bury the dead, and of the way to live as *one of us*. Without these ways there would be no corn, there would be no bread, there would be no life. These ways must be taught and they must be learned. They must be taught, some by the·fathers and some by the mothers; and it is the children who must learn them. But men are such that the stories which teach these ways cannot be shared unless they touch upon *the first things* and tell of a world which holds all our ways together. Without a whole, a unity, men are empty and life is without taste.

Unity meant a cause, and a cause meant a beginning, and a beginning meant a God or gods. But how united is the world? Did not the wind destroy what the sun had built? And if the Sun God was at war with the God of the Winds, what determined the outcome of that war and the terms of its battles? The world had become a world of consciously willing gods and necessity.

Then things began to happen. A traveler came to a village one day and found there a different way of baking bread and a different way of planting corn. The bread back home was better because it was *real bread,* but he could see with his own eyes that their corn was higher and fuller and richer than his corn.

Then too there was the half-breed, the man with two stories instead of one. For Strauss, we of the western world are such half-breeds, perhaps the bastards *par excellence.*

What holds our parents, Jerusalem and Athens, The Bible and Greek Philosophy, together and distinguishes them from myth, is that each of them

1 Lecture given 1 December 1990 at St. John's College, Santa Fe

came to grips with the problem of the corn and the bread. What holds them apart is the way in which they grasped that problem. Eventually, we shall have to ask ourselves whether that which holds them apart is not ultimately that which holds them together as well.

Of this moment when myth began to fall apart, Strauss wrote: "Unfortunately, the divine law, the *theos nomos,* to use the Greek image, leads to two fundamental alternatives: one is the character of Greek philosophy, the other is the character of the Bible." However much this statement may require revision in the course of our study, we must not forget that its first formulation begins with the word *unfortunately.*

Someone among the ancient Greeks, whose name I do not know, must have been the first to say: "There is a vast difference between the bread and the corn. It is we who prefer the bread because we have grown up with it; but there must be something about the corn itself which makes it grow best in the way in which it does." In some such way as this, *nature* was distinguished from *custom.* This man, or another like him, was then forced to ask the same question of the first things and of his way of life, of the just and of the unjust. Were they like the bread? or were they like the corn? Others joined him and some said one thing while others said another. Necessity was transformed into nature, while the willful gods either withered away, or were transformed into man.

To what extent did those who set down the text of the Bible divine or sniff out something of the consequences of the way taken by Greek philosophy? It is doubtful that anyone will ever have a clear answer to that question, but we can try to clarify for ourselves something of the way taken by the Bible.

To deny the distinction between the corn and the bread once a hint of that distinction is in the air, would be to assert that our God, who has given us a law concerning the bread dedicated for use in the temple worship, is the same god by whose command the corn and all other green things come forth from the earth. This would imply one God who is both Lord over the sun and Lord over the winds, and that He has chosen a particular people to be His people.

Mythic duality was replaced by an omnipotent God in the fullest sense. The willful gods became the one omnipotent God, and necessity fell under His command. As you may know, there is no Biblical word for nature; and it should be noted that even the medieval term *teva* comes from a word meaning *to dip.* The reference is to a signet ring. Nature is stamped into a thing from the outside; it does not grow from within the object itself. Man does not beget man nor dog beget dog because of its own nature, but in obedience to the command, "Go forth and multiply."

I believe that what struck Strauss most about the Bible is the depth of insight with which it presents what, for him, was the fundamental alternative to Greek philosophy. An omnipotent God was an essentially unknowable God. For Strauss, this consequence is fully expressed by the Bible in the verse which he translates as "I will be what I will be." For both Plato and the Bible, our daily lives are based on trust. You all trust that the sun will rise tomorrow morning, that your coats and hats will keep you warm, and that when you leave this room, your homes and the rest of the world will be out there waiting for you. For philosophy, this trust is based on the intrinsic intelligibility of what lies behind the world, whether that be as *nature,* or as *atoms and the void;* whereas for the Bible, our trust is in God's love and the covenant into which He has freely entered with us, first with Noah and through him with all living things, then with Abraham. While the argument between Abraham and God prior to the destruction of Sodom and Gomorrah shows that a pious man will have a care for God's justice in human terms, the binding of Isaac shows that God is not bound by His covenants in any humanly intelligible way, but that through His steadfast love He will freely keep them.

Much as these two ways, the Bible and Greek philosophy, stand as inverted images one of the other, there is one common agreement between them. Not riches, nor even courage, but *justice* is the highest human virtue. There is also this further agreement: that the right way is not immediately at hand but must be sought. There is, however, less agreement on where and how this search is to begin. For each way, this search is the most important act a man can undertake, and for neither way can it be undertaken directly. For each, there must be something underlying justice and supporting it. To this extent, we share the same conditions that led the earliest of men to look for a story that touched upon the *first things.* For the Bible, that underlying something is the *will of God,* which must be heard with humility and with awe; for Greek philosophy it is *theoria,* which must be looked into with boldness and with wonder.

This means inquiring into the world around us and into the nature of all things. But the Bible plays down the role of the world around us, and in particular the role of heaven. What for Plato is a visible god is for Genesis no more than a timepiece from which man can learn nothing more than the time of day unless he sees it as the handiwork of an all-powerful, all-loving God; whereas for philosophy, the discovery of nature essentially means that there is no conclusion to be drawn from the things that are made by man (and hence rely upon man) concerning the things that are not man-made but have within themselves their own source of motion and rest.

While God's warning to Cain, *If thou dost well...*, does seem to imply that there is available to man some sense of God's will from within himself, the remainder of the account shows only the inadequacy of this inner sense. While it may be enough to rouse a man to search out the word of God, it is not sufficient for human existence. Man is in need of divine revelation. In both traditions the goal of this search is called *wisdom*.

For the Bible, this simple sense of awe is most at hand in the desert and among the shepherds; but if man cannot or will not live that way, God is willing to allow him to live in a city, provided that it is a holy city ruled over by His anointed shepherd as king. Greek philosophy is at home in the *polis*. Its rise presupposes the leisure that only the city can provide. It is therefore willing to live with that certain amount of injustice which political life necessarily entails.

Where does this leave us? Mr. Strauss had a wonderfully simplistic way of putting it. *One believes in miracles, the other does not.* I believe that this intentionally bold and austere formulation of the question was meant to draw a clear and sharp line between what Strauss saw as the two fundamental alternatives for human existence. The reason for the harsh formulation is that much of western thought during the last two thousand years has been an attempt to harmonize our two roots and, in a sense, to return us to a time before that "unfortunate" discovery which led to the splitting up of myth into Greek philosophy on the one hand, and the Bible on the other.

For Strauss, this attempt was doomed to failure. As things looked to philosophy, the one thing needful for man—autonomous understanding—was the one thing denied him according to the Bible. That being the case, the only harmonization to be looked for would be a kind of handmaidenship. Either philosophy would be reduced to a method of deduction from received principles, or religion would be reduced to political exigency. In either case, neither would prosper; and the slave was sure to revolt.

Or one might try to raise oneself above the conflict. Needless to say, I am not referring to skepticism. Skepticism, in its own terms, must relate to others in their own terms. Skepticism, therefore, must be willing to place itself in the middle of the conflict. Rather I am speaking of what Strauss called by the term *contextualism*. Contextualism is the attempt to understand each thought in terms of where it was, and when it was. In so doing it prevents itself from coming to grips with the thought of the thinker as it was thought *by* the thinker; because what was of prime importance to the thinker cannot be of prime importance to the historicist. To see a thought as being important in its time and in its place is to deny it importance simply. But this would imply certain knowledge that the thought of the thinker was unimportant in itself; and that, of course, could only be established by returning to the plane of the conflict and again asking the question, *what is justice, how is it to be found*?

Part of what I mean by Strauss's "good-natured theory of history" is an account of history that does not abandon the plane of history. But what would it mean, then, this "returning to the plane of the conflict?" A great and totally meaningless battle would ensue, for each side argues its case in a court not recognized by the other. Spinoza's rational arguments against miracles presuppose reason. That is, they presuppose that there can be no miracles—the very thing which itself was to have been proved. Reason has acted unreasonably. In the same way, as Strauss goes on to argue, even Pascal's argument about the wager presupposes faith. Neither side can ever defeat the other, because each is wholly impervious to the weapons of the other.

There is one other way in which philosophy might win. If someone could give a completely satisfactory account of the whole, that would settle the matter. But, as Mr. Strauss was so fond of putting it, so far as we know that has never occurred. Note that there is no theoretical proof that it cannot happen, or that it will not ever happen. Thus, each future claim will have to be met; and we will be forced either to accept it or to articulate our doubts.

Now we seem to be at a standstill. We cannot withdraw, and we have no good reason for going one way rather than the other. The only solution, then, would seem to be simply to pick one side over the other, either at random or according to our natural bent. For Strauss, this would necessarily imply accepting the way of Jerusalem over the way of Athens. To him it was perfectly clear that if it was a matter of choice alone, then to have faith in faith was simply more honest than to have faith in reason.

Strauss was greatly moved by these considerations; and I believe that in spite of his natural sense of curiosity he would have accepted them simply, had it not been for Socrates—the man who knew that he did not know.

Perhaps we can understand that better by considering a passage from the *Meno*. As you remember, at first no one except Socrates believed that the boy knew how to find the square that is the double of a given square; but by the end of their discussion everyone except Socrates believed that the boy had come to know how to find the square. Instead, Socrates said:

> And at this moment those opinions have just been stirred up in him like a dream; but if he were repeatedly asked the same questions in a variety of forms, you know he will have in the end as exact understanding of them as anyone.

What was the boy missing? Let's first take a closer look at what he did see. He had seen that the square on the diagonal of a square is equal to *double* the square on the side. But let us reword the proposition only slightly and see what happens. It will now read: "The square on the diagonal of a square is equal to the *sum* of the squares on the other two sides"; of course the double is just a special instance of the sum—the sum of two equals.

Now we can begin to see what Socrates meant. The truth of the proposition has nothing at all to do with the equality of the two sides. It is simply the Pythagorean Theorem and would be true in any case. Of course, the boy does not know that yet; and it will require all the care and effort he has to discover it. But when he does, he will never again consider what he now possesses to be a piece of knowledge in itself. It will become only a minor case of a much larger proposition, scarcely worth mentioning in itself. For the present, the boy's horizons are small; but one day he might wonder to himself, "What would happen to what I have seen if the sides were no longer equal?" Or he might go on to ask himself, as Euclid once did, "What would happen if the angle were no longer a right angle?" If so, he would find that the square on any side of any triangle is equal to the sum of the squares on the other two sides *minus a certain rectangle* which, in the case of the right-angled triangle, has shrunk to nothing. What he has seen up to now, then, is only a part; and, as we have seen, it will disappear into a larger part as if it were a dream.

For Socrates, there are still so many ways in which the boy does not fully understand what he has done. It will not be until he tries to double the cube that he will begin to understand that his work with Socrates on that day was relatively simple only because he had limited himself to the surface. I believe that for Socrates, knowing the beguiling character of the theorem is itself part of knowing the theorem itself.

The Pythagorean Theorem, which gives the diagonal of a square, can be expressed by the equation $AB^2 + BC^2 = AC^2$. A similar, expanded equation, $AB^2 + BC^2 + CD^2 = AD^2$, gives the diagonal of a cube. Thus, if ever in my wanderings I should happen to meet an equation of the form

$$AB^2 + BC^2 + CD^2 + DE^2 = AE^2,$$

as once happened to Pappus, it might make me begin first to talk about, and then to wonder about, some kind of a supersolid; or, if we are talking about sensible objects, it might even tempt me to speak about a fourth dimension, as it did Minkowski.

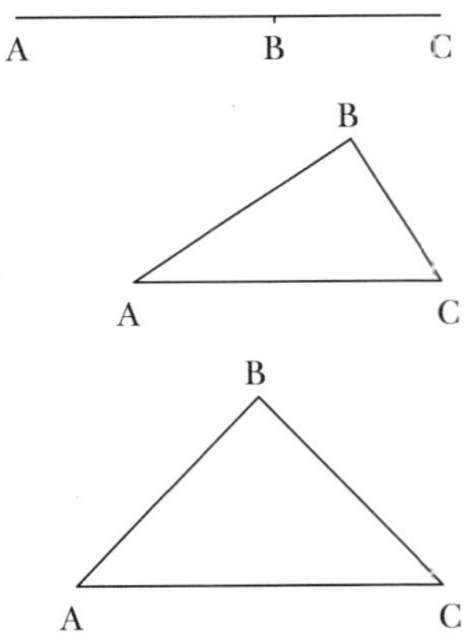

With this in mind, let us go back to the form $AB^2 + BC^2 = AC^2$. Except for one missing term, it calls to mind the form $AB^2 + 2AB,BC + BC^2 = AC^2$, which is the square on the straight line ABC.[2] Without that term, AC^2 is less than it was. The missing term therefore means to me that I have found a shorter path from A to C than by going straight through B; B must be off in some new dimension, the existence of which I was unaware of till now. The Pythagorean Theorem no longer means to me exactly what it meant before reading Minkowski. I do not mean to imply that Socrates had read Einstein: only that his formulation, his talk about seeing things "as if in a dream," seems to leave room for the fact that someday, someone might.

How does this relate to the double square? If in our diagram, we let AB = BC, the straight line ABC will be double AB; and the boy's first wild guess was that the double line gave the double square. Under further questioning he came to see that it was *AC the diagonal of the square* which gave the double square. What the boy did not see was the need to bend the straight line into a new dimension.

I sometimes wonder what would happen to our poor little theorem if it suddenly found itself in a world in which Euclid's fifth postulate was no longer within the horizon. Would the theorem simply disappear, or would it just transform itself again? Since it is not clear to me what *measurement* would or could mean under such conditions, I'm not sure that our theorem could survive. Socrates' contention, however, is that the possibility of anything even appearing to make sense within any horizon must be present within the grandest horizon.

Well, then; if, as Socrates says, to know something is to be able to recognize it wherever and whenever we happen to meet it, then perhaps I don't know the Pythagorean Theorem as I had thought I did, but have only seen it as if in a dream. Of course I can usually recognize it in my own freshman math class, and I think we have been able to spy it out even when it was hiding in the straight line, or when it turned up one day without its right angle.[3]

I also wonder what the boy will think of that day he spent with Socrates, when he learns the full Pythagorean Theorem and sees the limited nature of his earlier horizons—and hence, perhaps, of his present horizon as well. I hope that it will not lead him into despair. Strauss had a beautifully pithy way of putting it: *to see your own horizons is already to have seen beyond them.*

Although I have been speaking about the hypotenuse of a right triangle, I have partly been thinking of some days last summer which I spent reading *Mein Kampf.* It seemed important to me to try to see for myself what it was

---

2 $(AB + BC)^2 = AB^2 + 2AB ,BC + BC^2$.

3 Euclid, Book II, Props. 7, 12, 13.

within that horizon that so dazzled men's eyes that they never looked beyond that horizon's unspeakable narrowness. I also wanted to see whether I would have seen my own way had I been there. If Strauss is right and *thought leads beyond itself,* the way to burst out from within should have been there.

Such questions lead one to remember Socrates's account of his discussion with Thrasymachus. Thrasymachus saw a world in which justice was the advantage of the stronger. Socrates begins by asking him whether, if eating a great quantity of beef is to the advantage of Polydamas the pancratiast, that means it is to the advantage of all of us to do so. It was a silly question, not what Thrasymachus had meant at all. Not one word was used in the right way; but all the same it made Thrasymachus a bit nervous to see his words leading to a larger world than he had focused on. What Thrasymachus had meant, of course, is that justice is the advantage of the ruler, and that all must follow the ruler's command. As it turns out, somewhere on the periphery of his world was the notion that stronger knows what is truly to his own advantage and what is not. Somehow Thrasymachus must have always known that, or his world would never have made any sense to·him; but he could not focus on that notion without having to make a radical reinterpretation of his understanding of the world.

Once Thrasymachus had been forced to consider the environment of his own thought, however, he embraced it with vigor even though Socrates had given him the opportunity of rejecting it.

> "It doesn't make any difference, Polemarchus," I said; "if Thrasymachus says it that way now, let's accept it from him. Now tell me, Thrasymachus, was this what you wanted to say the *just* is: what seems to the stronger to be the advantage of the stronger, whether it is advantageous or not? Shall we assert that this is the way you mean it?"
>
> "Not in the least," he said. "Do you suppose that I call a man who makes mistakes 'stronger' at the moment when he is making mistakes?"
>
> "I did suppose you to mean·that," I said, "when you agreed that the rulers are not infallible but also make mistakes in some things." [340 c]

This step was fatal to Thrasymachus's argument, though it was hardly fatal to the man. It meant that Thrasymachus was forced to step outside his hero to see his wisdom rather than remain inside to feel his power. He does this partly because he has to make sure of his power in the future, but partly because he wants to be admired; and to be admired means to be admired from the outside. This means the he wants to be recognized as possessing an art. But when Socrates threatens to make Thrasymachus step beyond his art, as those who care for horses are judged by those who can tell whether a horse has been well cared for or not, Thrasymachus turns on him:

> "Tell me, Socrates, do you have a wet nurse?" "Why this?" I said, "shouldn't you answer instead of asking such things?" "Because," he said, "you know she neglects your sniveling nose and doesn't give it the wiping you need; since it is her fault you do not even recognize sheep or shepherd." [343 a]

Shepherds, according to Thrasymachus, only rule for the good of the sheep in so far as it makes them fatter and gives the shepherd more meat. Stepping outside in order to see what is best for the sheep themselves can at best only be ancillary. The only real question is *what is best for the shepherd.* Thus Thrasymachus has escaped Socrates; but in so doing, he has failed to escape himself.

Near the end of their discussion, Thrasymachus readily agrees to the statement:

> "Let us say it, then, as follows," I said. "The just man does not try to better what is like but what is unlike, while the unjust man tries to better both like and unlike?" [349d]

For example, the musical man is able to best the unmusical man precisely by tuning his lyre in the same way that other musical men do, not by bettering them; and the same is true of the wise and of the just. But unlike the artisans, the unjust man who wishes to better—that is, to better anybody or anything—can have no common goals outside himself in terms of which he could be praised. Thrasymachus, who so wanted to be seen and heard has finally been pried out of himself in order to see whether he could be seen. He blushes and remains as silent as one who has been seen by a bear.

Socrates's goal, then, is not to defeat Thrasymachus or to prove that he is wrong. Perhaps that can't even be done. What he may be able to do is to help him find those bits on the periphery of his vision which might lead him beyond his own horizon.

The case of Meno is a different matter. Meno has a teaching, or a dogma—one might even say an ideology. But it is only the dead remains of a thought which first grew up in another mind, the mind of Gorgias. People like Meno do not speak with their own voice. Their voice is not home-grown, because it did not arise from within their own horizons. Our own voice is the only voice able to articulate all those blurred and obscured thoughts, which are first stirred up from within our own particular partial awareness of the surface of things as it lies within our own particular shaggy and ill-defined horizon. Other thoughts may have a brilliant central focus; but when they have been poured from one mind into another, they lose that particular periphery which once connected them to the land in which they were born. For this reason there are no pathways which could lead a student back to his own horizons and beyond. Too often, what is a living thought for the teacher becomes a hardened dogma for the student—precisely because he cannot reach out to the horizon but is continually forced to return to the center, which he can repeat in comfortable repetition.

We can now restate Stauss' difference from, and objection to, historicism. Historicism sees itself as giving a scientific account of the thoughts of the past in terms of the horizons within which those thoughts were thought. That scientific account presupposes a closed and well defined horizon, in terms of which the activity of thought within that horizon can be made intelligible. But for Strauss, no partial horizon is well defined. Hence, it cannot be considered closed, no matter how closed it may be to a man like Meno.

For Strauss it was more a matter of how wide or how narrow our horizons are, how deep we are willing to look, and how willing we are to look at their shaggy edges and hence to look beyond them. It is also a question of whether we have such a nature as to thrive *entre les deux infinites* or whether we find it a frightful land from which we must escape today.

For Strauss, the existence of a realm that lay between knowledge and ignorance meant that there was a land in which inquiry could be pursued with honesty. This implies that, while philosophy can proceed only by examining the way things seem to be, as long as the horizons are open, it must learn to coexist with theology. Only theology, he thought, can prevent philosophy from mistaking itself for *sophia;* and such presumed *sophia* had no other choice than to express itself as ideology, which would soon congeal itself into tyranny. I also believe he thought that only philosophy was in a position to return the favor. It is in connection with this conclusion that he makes it clear that, for both philosophy and theology, the love of justice must be prior even to love of country, if love of country is to have any meaning at all.

The Broken Staff[1]

Shakespeare's last play begins at sea in a wild and chilling storm. We hear the thunder and see the lightning. The sea-spray wets our cheeks and weighted clothing hangs limply about us. Yet there was no storm; and anyone in the audience not taken up by the magic clearly sees that our clothing is as fresh as the moment we stepped out of the house to go to the theater.

The Master of the Ship is the first to speak:

> Boatswain! [I.i.1]

The Boatswain answers:

> Here, Master; what cheer? [I.i.2]

The Master speaks only to the Boatswain, and it is he who is to give to the men the words of the Master. No one else ever hears him. In a manner that begins to remind us of the Lord speaking to Moses from the top of Mount Sinai, the Master says:

> Good; speak to th' mariners. Fall to't yarly,
> or we run ourselves aground. Bester, bester. [I.i.3–5]

Then the Master, like the Lord of the Christian Old Testament, disappears, causing both the King and the Duke to demand:

> Where's the Master? [I.i.10, 13]

When the Boatswain answers

> Do you not hear him? [I.i.14]

we can only think of Moses and his constant reminder, "Hear the word of the Lord!" Again the word can be heard through the mouth of the spokesman. But unlike the Lord, who led His people into the Land of Israel, the Master is more at home in the open sea. He fears dry land.

The Boatswain, daring to the sea, and fearful of the land, has a hearty cheer for his men:

> High, my hearts! Cheerly, cheerly, my hearts! Yare, yare!
> Take in the topsail, tend to the master's whistle, blow till
> thou burst thy wind, if room enough. [I.i.6–9]

The last phrase means that as long as the storm does not send him too close to the shore, that is, as long as it gives him enough of what sailors call "sea

1 Lecture given 23 September, 1983 at St. John's College, Santa Fe

room," it can blow its worst. His orders are flawless, but his command over nature leaves him oblivious to mere convention and thus to all those things that render political life possible. To the royal party itself he says:

> Hence! What care these roarers for the name of king?
> To cabin! Silence! trouble us not. [I.i.17–19]

Gonzalo, the honest old Counsellor, for whom I hope we shall learn to have high respect and love, in spite of the fact that we shall be forced to go well beyond him, reflects:

> I have great comfort from this fellow. Methinks he hath no drowning mark upon him; his complexion is perfect gallows. Stand fast, good fate, to his hanging; make the rope of his destiny our cable, for our own doth little advantage. If he be not born to be hang'd, our case is miserable. [I.i.30–35]

We can well understand Gonzalo. He is thinking of men like Theseus, David, or Wyatt Earp, those who can slay giants or rid the seas of pirates. Without them civil life cannot exist; and yet they themselves can never quite find their way back into that life. There will always be an Ariadne or a Bath-Sheba. How can a man who has shown his excellence in that world which knows no law ever place himself under the external rule of other men? Ancient wisdom, both biblical and Greek, tends to show that those very men who are so needed for the establishment of political life cannot themselves be contained within its borders and lead to its ultimate destruction. Yet in the final scene of *The Tempest*, we shall meet the Boatswain again. This time he will indeed be on dry land; but instead of being hanged, he will say:

> The best news is that we have safely found our king and company...
> [V.i.121–122]

What has Gonzalo, noble as he is, overlooked that has made this change possible? Part of our answer may lie in his final speech of the act:

> Now would I have a thousand furlongs of sea for an acre of barren ground, long heath, brown [furze], anything. The wills above be done!
> But I would fain die a dry death. [I.i.69–72]

Gonzalo is as lost at sea as the Boatswain is on land. Weaving them together will be part of Prospero's art.

We first see Prospero's art through the eyes of his daughter, Miranda. We see what her eyes see, we do not see what they do not see. From her we learn that his is a violent art, which does not look for the mean but works by the direct confrontation of opposites.

> If by your art, my dearest father, *you have put the wild waters in this roar, allay them. The sky, it seems, would pour down stinking pitch, but that the sea, mounting to th' welkin's cheek,* dashes the fire out. [I.ii.1–5]

But we learn more. This is the first opportunity Miranda has had to demonstrate concern for others. The storm has served as part of her education, and she passed through it well, though there is much she still does not understand. In her belief that all perished in the storm, she said:

> Had I been any god of power, I would have sunk the sea within the earth. [I.ii.10–11]

These lines contain a remarkable and deep mixture of seeing and not seeing. No one has been harmed, yet if any harm had been done, her passion would have been right. Will it require a god? Prospero, as we shall see with greater and greater clarity, is not a god; though it may be well for her to begin by looking for a god. Then, too, although the total engulfment of the sea within the confines of the earth may not be the best way of weaving sea and earth together, Miranda has vaguely seen the problem raised by Gonzalo and the Boatswain.

◆

Miranda's next lesson is a lesson in history. It is a very hard lesson, and it contains more than history.

> Lend me a hand,

says Prospero,

> and pluck my magic garment from me. [I.ii.23–24]

This lesson cannot be given by magic, and Miranda must come to know that. Thus we see, even this early in the play, that there is a time when magic must come to an end. Prospero had carefully prepared Miranda for this moment by often beginning to tell her the story, but always concluding:

> Stay, not yet. [I.ii.36]

Prospero's non-magical beginning is to force Miranda to dig deeply into her own memory. She found little, but in looking she had contacted herself and was ready to go on. Old Gonzalo had once made the same appeal to the Boatswain:

> Good, yet remember whom thou hast aboard. [I.i.21]

But the only answer was:

> None that I love more than myself. [I.i.22]

Gonzalo had forgotten that in the ever-present sea there is no past and no memories.

From her father Prospero, Miranda, living alone with him and one servant on a desert island, learns that her father had once been the Duke of Milan, but his brother Antonio, with the help of Alonzo, king of Naples, stole the dukedom, putting Prospero and his daughter out to sea in an old leaky boat. Gonzalo was in on the plot, but, being a noble man, he saw to it that Prospero was provided with clothing and with his books.

Miranda thinks that she has listened well to Prospero's story, and in a way she has; but Prospero does not think so. He continually complains that she is not listening. Perhaps he thinks that she did not fully attend to the words:

> I pray thee, mark me. I, thus neglecting worldly ends, all dedicated to
> closeness and the bettering of my mind with that which, but by being
> so retir'd, o'er-priz'd all popular rate, in my false brother awak'd
> an evil nature; and my trust, like a good parent, did beget of him
> a falsehood, in its contrary as great as my trust was; which had
> indeed no limit, as confidence sans bound. [I.ii.88–96]

Perhaps she did not hear him say that the heart of the problem lay not with Antonio, but in his own lack of political action. Perhaps he had read in the *Republic* that justice would never come to be until kings became philosophers, or philosophers became kings; but such thoughts can come to a man as an immediate call to action only before Plato's abortive attempt to bring justice to Sicily. After the fiasco in Sicily all such seemed to have the status of an eternal dream. Good men had better stay out of politics; they can do no good, and can only become corrupted themselves. But now, on the island, political thoughts seem to have returned to Prospero, some two thousand years after Sicily. How? What has happened? We must know more about Prospero and his magic. After telling her the story, Prospero again dons his cloak and gently puts Miranda to sleep.·

While Miranda sleeps, let us meet the chief instrument of Prospero s magic, Ariel. Indeed we meet him as a pure means to another's end, though in himself he seems to be completely lacking in any desire other than an all-consuming desire for total liberty. Who is he, and how has such an inversion taken place in him? Some have called Ariel the spirit of philosophy, others the spirit of poetic imagination, or poetic inspiration. But all these things seem too narrow. Let us say no more than that he is whatever it is that pulls man beyond the physical needs of himself or of others, whether it be to sing or to dance or to play or to think.

It was Ariel's task to perform this great tempest while making sure that in fact not a hair perished. After his tasks are completed, Ariel begins in turn to give a history lesson to Prospero; and again it is given through recollection:

Ariel: Let me remember thee what thou hast promis'd, which
is not yet perform'd me.
Prospero: How now? Moody? What is't thou canst demand?
Ariel: My liberty. [I.ii.242–245]

Prospero's playful cajoling and his loving promise of freedom are usually enough to win over the good natured Ariel; but this time he is forced to return the history lesson:

Hast thou forgot the foul witch Sycorax,
who with age and envy was grown into a hoop? [I.ii.257]

In the days before Prospero had come to the island, there was another worker of magic, one Sycorax, who had enslaved Ariel and forced him to do her bidding. When Ariel, out of his own sense of delicacy, would not act upon her commands, she confined him in a cloven pine from which she could not free him. Later she died, leaving him in bondage. Ariel must be reminded once a month that if he is unwilling to work for a noble master, he will be captured by an ignoble master. With this, the talk ends. But we are left with a question. How shall we tell the difference between the magic of Prospero and the rnagic of Sycorax?

When Ariel leaves and Miranda wakes, we meet Caliban, a savage and deformed slave. He comes in cursing, laying two charges against Prospero: one, that Prospero has taken the island from him by force; the other, that Prospero's great gift of language has taught him nothing but to curse. Caliban's claim always makes us just a bit uncomfortable, since there is something just and right about it.

This island was mine by Sycorax my mother, which thou tak'st
from me. When thou cam'st first, thou strok'st me and made much
of me, would'st give me water with berries in it, and teach me how
to name the bigger light, and how the less, that burn by day and
night; and then I loved thee and showed thee all the qualities o'th' isle,
the fresh springs, brine-pits, barren place and fertile. Cursed be I
that did so! All the charms of Sycorax—toads, beetles, bats, light
on you! For I am all the subjects that you have, which first was
mine own king; and here you sty me in this hard rock, whiles
you do keep from me the rest o'th' island. [I.i.331–344]

Throughout the play, Shakespeare gives us just enough insight into Caliban to make us wonder what he might have been like had Prospero not come along. At the same time, we see that Caliban could never come to understand why Miranda is not for him, and that will lead to bitterness and ugliness. As we shall see later, there is one thing Prospero could have done to get Caliban to abandon his desire for Miranda and thereby bring peace to his island, but that is the one thing that Prospero is determined not to do.

But it is now time for Miranda to meet Ferdinand, the son of King Alonzo. He enters the stage walking around in a kind of daze. Ariel sings him two songs. Each time his reaction is the same: he *remembers* something. As we know by now, that is a good sign. One day he will make a fine husband for Miranda, though not yet. Ariel's second song goes like this:

> Full fathom five thy father lies; of his bones are coral made; those
> are pearls that were his eyes: nothing of him that doth fade but doth
> suffer a sea-change into something rich and strange. Sea-nymphs
> hourly ring his knell: Ding-dong. Hark! Now I hear them, ding-dong bell.
> [I.ii.397–405]

Ferdinand then says:

> This ditty does remember my drown'd father. [I.ii.405]

Ferdinand speaks more truly than he knows. The next time he sees his father, the king will be much changed. Alonso is a fine example of the effects of Prospero's art. If we could only understand the lines

> Nothing of him that doth fade but doth suffer a
> sea-change into something rich and strange, [I.ii.399–401]

we night be able to begin to understand the difference between the old and the new magic.

When Miranda first sees Ferdinand, she thinks that he is divine; and when he first sees her, he takes her to be a goddess. They are, of course, both deeply wrong. Each is very human, and they can never be in love till they understand that. However, falling in love may be another matter. They must learn to love each other as human beings, but falling in love may require the almost momentary divinity of magic. The rest of Act I is devoted to Prospero's reduction of Ferdinand to the human.

Though some parts of the next scene have a kind of clarity about them, others remain most obscure; and much of what we have to say regarding it must be taken as at best no more than a tentative approach.

Gonzalo begins by offering solace to the king, who believes. that his son Ferdinand was drowned in the tempest. He is not, however, so indelicate as to mention the word *death,* but speaks only of "our hint of woe." In effect he says that wisdom lies in the realization that there is not much man can do against the sea; one can only be thankful for whatever good remains. This view of life, which we had already seen in Act I, may account for his willingness to participate in the abduction of Prospero twelve years earlier. For him, the best one can do is to make sure that the right book gets into the right hands.

The king's brother, Sebastian, a friend of Antonio, then accuses Gonzalo (whom he will later refer to as Sir Prudence) of thinking like a clock, that is to say, of relying only on prudence and lacking imagination. We know only too well what Sebastian means by that. In him, imagination is constantly working out schemes of self-interest narrowly understood. But properly understood, he may have a point. This leads to a disagreement about the color of the island. Gonzalo thinks it is green, Sebastian and Antonio think it tawny. This, in turn, gives rise to the following exchange:

> Gon.: Methinks our garments are now as fresh as when we put
> them on first in Afric, at the marriage of the king's fair daughter
> Claribel to the king of Tunis.
> Seb.: 'Twas a sweet marriage, and we prosper well in our return.
> Adr.: Tunis was never grac'd before with such a paragon to their queen.
> Gon.: Not since widow Dido's time.
> Ant.: Widow! A pox o'that! how came that widow in? Widow Dido!
> Seb.: What if he had said "widower Aeneas" too? Good Lord, how
> you take it!
> Adr.: "Widow Dido," said you? You make me a study of that. She
> was of Carthage, not of Tunis.
> Gon.: This Tunis, sir, was of Carthage.
> Adr.: Carthage?
> Gon.: I assure you, Carthage.
> Ant.: His word is more than the miraculous harp.
> Seb.: He hath rais'd the wall and houses too.
> Ant.: What impossible matter will he make easy next?
> Seb.: I think he will carry this island home in his pocket and
> give it his son for an apple.
> Ant.: And sowing the kernels of it in the sea, bring forth more islands.
> [II.i.67–94]

There things begin to get a bit foggy. It is clear why Gonzalo, who, as we remember, has a great fear of the sea, is surprised that the sea should have been so kind; but it is not clear why he should insist so strongly on the relation between Tunis and Carthage, or between Dido and Claribel. Nor is it clear why this leads Antonio to charge him with utopianism:

> His word is more than the miraculous harp. [II.i.85]

My best guess is that Sebastian and Antonio think that Gonzalo is suggesting that if the death of Dido was, as it were, the original sin behind the fall of Rome, then by returning Dido in the form of Claribel, Alonzo has paid the debt. Sebastian will have none of this, and he accuses Alonzo of being the true cause of their present troubles by marrying Claribel to an African. Alonzo accepts that and for the first time feels the weight of a past action. Through Antonio, Prospero's magic is at work. But Gonzalo accuses Antonio of speaking in malice. Even if what he says is true, this is not the time to add to Alonzo's suffering. One of the points of utopianism is to bring a plaster to the sore.

Gonzalo then gives his great utopian speech, extolling the wonders and beauties of primitive life. It begins:

> Had I plantation of this isle, my lord—
> ... and were the king on't...

and ends

> but innocent and pure; no sovereignty. [II.i.142–153]

As Sebastian is only too quick to point out, utopianism is self-contradictory. At best it can only lead us to a deeper understanding of why things are as they are; it has little power to change them. Gonzalo then answers:

> You are gentlemen of brave metal; you would lift the moon out of her sphere, if she would continue in it five weeks without changing.
> [II.i.182–184]

For Gonzalo, the only alternative to utopianism is the conquest of nature; but nature is too uncertain and hence cannot be conquered.

At this point, Ariel enters and puts the others to sleep, allowing Antonio to seduce Sebastian into the murder of Alonzo and Gonzalo. We must limit our remarks to the fact that the turning point seems to be when Sebastian says:

> I do remember you did supplant your brother Prospero. [II.i.270]

A good memory turns out not to be good simply; it still depends upon who remembers what and when. Perhaps the best thing that can be said is that what pops into memory unbidden may reveal character best.

◆

The next scene begins with Caliban alone on stage, cursing Prospero for all his pains and aches. These lines are Shakespeare at his best; in them we can see both Caliban's gentility and Prospero's wisdom at the same time, though they seem to contradict each other.

The next character to arrive on stage is Trinculo. Caliban falls flat and hides under his cloak, hoping to be ignored. Just as the world is green for Gonzalo and tawny for Antonio, for Trinculo it is wet.

> That same black cloud, yond huge one, looks
> like a foul bombard that would shed his liquor. [II.ii.20–21]

In fact, Trinculo ends up climbing in under the cloak with Caliban (whom he takes to be a native struck by lightning) just to avoid the downpour that no one else ever sees or speaks of. Nor can he always tell the difference between a man and a fish, even when it's *legg'd like a man,* and has *fins like arms.* Once, when asked how he came ashore, he said:

> Swam ashore, man, like a duck. I can swim like
> a duck, I'll be sworn. [II.ii.132–133]

In short, he seems to be a kind of cosmic version of our friend the Boatswain; and we must remember that when the mariners cried

> All lost! To prayers, to prayers, all lost!

the Boatswain merely remarked

> What, must our mouths be cold? [I.i.55–56]

Trinculo does not think much of Caliban, either. In the course of his last ten lines, he uses the word *monster* eleven times. Each use is accompanied by its own adjective. The list includes such words as *weak, shallow, howling, drunken, perfidious, puppy-headed, abominable,* and *scurvy*; but above all, *most poor credulous monster*. In contrast, his friend Stephano only uses the word directly of Caliban once. The last line of the act reads:

> O brave monster! Lead the way. [II.ii.192]

But we haven't met Caliban yet, so let us go back and do so.

After Trinculo had crept under the cloak with Caliban, Stephano came staggering in singing a scurvy tune. He saw what he took to be a two-headed monster with four legs, but he was not afraid of it. He had saved a whole butt of wine and gives some to one of the heads. In fact it turns out to be Caliban's. The head then says:

> That's a brave god and bears celestial liquor.
> I will kneel to him. [II.ii.121–122]

Stephano has his wine in a bottle that he made with his own hand from the bark of a tree, and which he offers to Caliban again, saying:

> Come, kiss the book. [II.ii.145]

I believe we are to take Caliban's reply as both heartrending and contemptible:

> I prithee, be my god. [II.ii.152]

At any rate, that is what brings forth all those *monster* remarks from Trinculo. Caliban, on the other hand, makes what I find to be a moving speech to his god:

> I prithee, let me bring thee where crabs grow; and I with my long
> nails will dig thee pig-nuts, show thee a jay's nest, and instruct thee
> how to snare the nimble marmoset. I'll bring thee to clust'ring
> filberts, and sometimes I'll get thee young scamels from the rock.
> Wilt thou go with me? [II.ii.171–176]

But as they leave he goes out singing:

'Ban,'Ban, Cal-Caliban
has a new master, get a new man.
Freedom! High-day! High-day! Freedom!
Freedom! High-day! Freedom! [II.ii.188–191]

For Caliban, to be free is to have a new master, a god!

◆

Act III begins by taking up the same subject, freedom and slavery; but this time from the point of view of love. As part of his humanization of love, Prospero has imposed upon Ferdinand Caliban's job of log-man. The action is a kind of see-saw, in which Miranda learns to question law and Ferdinand learns to obey it. Ferdinand is willing to accept what he will later call his *wooden slavery* on account of his love for Miranda.

The term *wooden slavery* not only reminds us of the logs; it also must remind us of Ariel's wooden slavery and Caliban's slavery to the wooden bottle and hence—*Come kiss the book*—to books. All these Ferdinand will accept for Miranda.

She, on the other hand, out of compassion, first tries to persuade him to rest, although she is fully aware of her father's command. Then she goes as far as to wish that lightning had burnt the logs. Then, in a manner that even reminds us of Antonio's argument, she says:

My father is now hard at study; pray now, rest yourself. [ ]

Later she tries to persuade Ferdinand to let her carry the logs, and finally she actually breaks her father's behest and reveals her name to Ferdinand. Prospero, who was secretly looking on, said only:

Heaven rain grace upon that which breeds between 'em. [ ]

We must now return to our comedians, who are no longer quite so comical. By means of his book, that is to say, his bottle, Stephano has completely succeeded where Prospero had failed. Caliban has become a dutiful subject. He has given up all claim to Miranda, whom he will win for his god by delivering up Prospero to be killed.

But Ariel, who has been present but invisible, begins to lure them with tabor and pipe. Shakespeare then does a very strange thing. He chooses this moment to give Caliban some of the most beautiful lines in the play. Let me read them to you:

Be not afeard. The isle is full of noises, sounds and sweet airs,
that give delight and hurt not. Sometimes a thousand twangling
instruments will hum about mine ears, and sometime voices that,
if I then had wak'd after long sleep, will make me sleep again; and
then, in dreaming, the clouds methought would open and show

riches ready to drop upon me, that, when I wak'd, I cried to dream again. [III.ii.143–152]

Shakespeare's world is a crazy multicolored world. We may have to leave Caliban one day, or even beat him, but Shakespeare also wants us to know what we will have lost. But Shakespeare will not let us rest with that. Ariel describes the last scene in the following words:

> Then I beat my tabor; at which, like unback'd colts, they prick'd up their ears and *advanc'd* their eyelids, lifted up their noses as they smelt music. [IV.i.175–179]

We have been duped, and Shakespeare has sent us napping, so that we might catch ourselves taking for beauty what has become bent over like a hoop. We have not noticed that, though Caliban may once have had a certain charm, the son of the old magic is the one of whom it will be said:

> and as with age his body uglier grows, so his mind cankers. [IV.i.291–292]

We must now return to the king's party. Old Gonzalo is tired, and Alonzo has given up all hope. Strange figures then appear, bringing in a banquet. With gentle actions they invite everyone to come eat, then they depart.

All are amazed—Gonzalo, too, but in a different way from the others. He wonders at their gentility and kindness; but to him they are certainly people of the island. Gonzalo is not in need of magic, and so for him there is no magic. Suddenly the banquet disappears, and in its place stands Ariel dressed as a harpy. He says, in part:

> You are three men of sin, whom destiny, that hath to instrument this lower world and what is in't, the never-surfeited sea hath caus'd to belch up you; []

and later in the same speech,

> But remember—for that's my business to you—that you three from Milan did supplant good Prospero; expos'd unto the sea, which hath requit it, him and his innocent child; for which foul deed the powers delaying, not forgetting, have incens'd the seas and shores, yea, all the creatures, against your peace. Thee of thy son, Alonzo, they have bereft; and do pronounce by me ling'ring perdition, worse than any death can be at once, shall step by step attend you and your ways; whose wraths to guard you from—which here, in this most desolate isle, else falls upon your heads—is nothing but heart's sorrow and a clear life ensuing. [III.iii.67–82]

In each case it is the sea that has revolted and takes its revenge.

After the harpy vanishes and the table is gone, Gonzalo remembers nothing; but Alonzo is shaken. He remembers only a singing of the wind and the lone word *Prospero*; but that is enough to cause him to remember his own guilt as if from his own memory and to see its relation to the drowning of his son.

Prospero then goes to visit the young ones. To Ferdinand he says:

> I have given you here a third of mine own life or that
> for which I live; [IV.i.2–5]

We also remember that when he was speaking to Miranda as they stood together, looking down on the wreck of the ship, Prospero had said to her:

> I have done nothing but in care of thee, of thee
> my dear one, thee my daughter, [I.ii.16–17]

We conclude that his magic was no more than one-third of Prospero's life. We further suspect that it was the whole of Sycorax's, and therein the difference may lie.

Ariel enters and, with Prospero, presents *some vanity of his art.* It is a play with three spirits, Iris, Ceres, and Juno. Iris, who describes and praises the richness of Ceres' bountifulness, invites her in the name of Juno the Queen to come to the grass-plot. Ceres answers with equal grace, but wants to know why she has been summoned. When she hears that it is to celebrate a contract of true love, she has some reservations. For she has forsworn the company of Venus and her blind boy because he plotted the means by which *dusky Dis* had got her daughter. But she agrees to come when she hears that

> Mars' hot minion is return'd again; her waspish-headed son
> has broke his arrows, swears he will shoot no more, but play
> with sparrows and be a boy right out. [IV.i.98–101]

The play presents the final end of the struggle between Venus and Ceres, a conflict which has kept the world apart for over two millennia.

Ferdinand and Prospero converse, and Ferdinand says:

> Let me live here ever; [IV.i.122]

Instead of telling Ferdinand directly that his wish cannot be granted, Prospero says:

> Sweet now, silence! Juno and Ceres whisper seriously, there's
> something else to do; hush and be mute, or else our spell is
> marr'd. [IV.i.124–127]

Then Iris tells us what this serious business is:

> You nymphs, call'd naiads, of the wand'ring brooks, with your
> sedg'd crowns and ever-harmless looks, leave your crisp channels,
> and on this green land answer your summons; Juno does command.
> Come, temperate nymphs, and help to celebrate a contract of true
> love; be not too late. [IV.i.128–133]

Again we see, as in the case of the boatswain, that the answer lies in making room on the land for the men of the water. If this is not done, all will have been in vain. Again we are assured that the sea-beings can become *ever-harmless*.

As Iris continues to speak, nymphs and reapers come and dance. Suddenly Prospero remembers the conspiracy on his life; and he himself mars the spell by crying:

> Well done! Avoid. No more! [VI.i.42]

The children are upset, but Prospero comforts them by saying:

> You do look, my son, in a mov'd sort, as if you were dismay'd. Be cheerful, sir. Our revels now are ended. These our actors, as I foretold you, were all spirits, and are melted into air, into thin air; and, like the baseless fabric of this vision, the cloud-capp'd towers, the gorgeous palaces, the solemn temples, the great globe itself, yea, all which it inherit, shall dissolve and, like this insubstantial pageant faded, leave not a rack behind. We are such stuff as dreams are made on, and our little life is rounded with a sleep. [IV.i.146–159]

For the first time, Shakespeare begins to draw the analogy between his own art and the art of Prospero. We shall see the analogy grow throughout the next act. We can now understand the implications of Caliban's opening of Act II:

> His spirits hear me, and yet I needs must curse. But they'll
> nor pinch, fright me with urchin-shows, pitch me i'th' mire,
> nor lead me, like a firebrand, in the dark out of my way,
> unless he bid 'em; [II.ii.3–7]

That can be said of Shakespeare, but it cannot be said of Plato. The philosopher, precisely because he writes for the few, must rely upon interpreters and popularizers to affect the world. But they can pinch and fright, and pitch into the mire, those whom Plato himself would not have treated so. But perhaps there is a way in which the philosopher can become a king: perhaps he must become a poet. The poet writes for a much larger audience, and in that sense may be said to rival even the Bible. But if the poet is to avoid the problems raised by Plato in the Ion, he must write in such a way as to enlighten and set free rather than to mystify and hold in bondage. But is not magic the only tool the poet has at his command? Such are the problems that led Shakespeare to reconsider the Republic two thousand years after Sicily.

But how can the poet be a king? Is he not rather a king of nothing? His palaces and temples are all made of cloth and go back into their boxes at night. When the show is over the actors go home, but where do the characters go? Even the theater, and the racks that hold the shoes and the hats that go on the men and boys that play the kings and fools—one day they will all be gone. Surely we will be forced to reconsider the problem eventually.

◆

Shortly before meeting Caliban, Prospero turned to Ariel and said:

> That was well done, my bird. [IV.i.184]

And again, after the encounter, he said:

> Shortly shall all my labours end, and thou shalt have the air
> at freedom. For a little, follow, and do me service. [VI.i.265–267]

Ariel's freedom will be mentioned again no less than four times in Act V, and it emerges as the second goal of Prospero's project. It is tempting to think that we have here found the other two-thirds of Prospero's life—the political, the sub-political, and the realm above the political. At any rate, at the end of the play, Prospero says of Caliban:

> This thing of darkness I acknowledge mine. [V.i.275]

At this point in the play, however, he can only say:

> A devil, a born devil, on whose nature nurture can never
> stick; on whom my pains, humanely taken, all, all lost, quite
> lost; and as with age his body uglier grows, so his mind cankers.
> I will plague them all, even to roaring. [IV.i.188–193]

The only thing that Prospero can do for Caliban is show him that his god is beneath him. To distract the would-be murderers from their goal, Ariel lays the court clothing out under a tree. Stephano and Trinculo fight over them; but Caliban, who wants to get on with the murder, looks on at his god in disgust. Being un-nurtured, he is at least not attracted by gaudy clothing.

All these threads come together in Act V. We pick them up with Prospero's great soliloquy:

> Ye elves of hills, brooks , standing lakes, and groves, and ye that on
> the sands with printless foot do chase the ebbing Neptune, and do
> fly him when he comes back; you demi-puppets that by moonshine
> do the green sour ringlets make, whereof the ewe not bites; and you
> whose pastime is to make midnight mushrooms, that rejoice to hear
> the solemn curfew; by whose aid, weak masters though ye be, I have
> bedimm'd the noontide sun, call'd forth the mutinous winds, and
> 'twixt the green sea and the azur'd vault set roaring war; to the
> dread rattling thunder have I given fire, and rifted Jove's stout oak
> with his own bolt; the strong-bas'd promontory have I made shake,
> and by the spurs pluck'd up the pine and cedar; graves at my command
> have wak'd their sleepers, op'd, and let 'em forth by my so potent art.
> [V.i.33–50]

Much of this speech is taken directly from Ovid, though Shakespeare shows himself perfectly willing to change whatever had to be changed. But perhaps we can see some of the reasons why Shakespeare may have chosen it. First, it is addressed to those in whom we have been most interested in all the time—all who are willing to flirt with Neptune, though they do not know how to live with him. It is also addressed to those *demi-puppets who by moonshine make the green sour ringlets,* the grass-plot where Ceres and Juno can meet.

With the help of these people, whom he calls *weak masters* because, although they are ready to live in a new world, they cannot bring it into being, he has *bedimm'd the noontide sun, and called forth the mutinous winds.* Caliban once said:

> All the infections that the sun sucks up from bogs,
> fens, flats light on Prospero. [II.ii.1]

In this play even the sun, that by whose light we see all things, can become infected and must for a time be bedimmed. He has *called forth the mutinous wind.* Prospero had once turned to Ariel to say:

> Hast thou, which art but air, a touch, a feeling of their afflictions…
> [V.i.21–22]

We therefore know that under the reign of the foul witch Sycorax, that air had become a *mutinous wind,* which she confined into a cloven pine. Throughout the play we have seen Prospero call forth that mutinous wind.

The war between the *green sea and the azur'd vault,* described so well by Miranda in Act I, was the condition under which we came to recognize both the virtues of the Boatswain and the need for reconciling him to Gonzalo. With regard to the line about *Jove's stout oak,* we have already seen the relation between books and trees, and why the new status of the written word requires a reconsideration of the Platonic position.

> Graves at my command have wak'd their sleepers,
> op'd and let 'em forth by my most potent art. [V.i.48–50]

And indeed they have done—the graves of Caesar, of Antony, and of the kings of England from John to Henry. Prospero began with a mixture of reminding and telling Miranda who she had been. The History Plays give us a past. But as Shakespeare points out in the Prologue to Henry the Fifth, he cannot *cram within the wooden O the very casque that did affright the air at Agincourt,* but we must *piece out his imperfections with our thought.* Like Miranda, we cannot learn from Shakespeare unless we are willing to strain within ourselves for the way.

In the History Plays, Shakespeare presents us with a past which we can become part of, as it becomes a part of us. It is a past full of those that we can admire, or laugh at, or fear, or love. Through them we begin to learn the meaning of words like Courage, Honor, Nobility, and their opposites. Our new past not only establishes for us a way of life, it al so gives us a language in which to think about our fellow men. Each of them begins to have an underived character in himself. If to be a king in the highest sense is to lay down his country's foundations and establish its ways, then perhaps Shakespeare was the only true king that England has ever known.

I should like to suggest that innate character, and hence the possibility of the noble, is the grass-plot upon which Gonzalo and the Boatswain can meet.

But Prospero's speech goes on:

> But this rough magic I here abjure, and, when I have requir'd some heavenly music, which even now I do, to work mine end upon their senses that this airy charm is for, I'll break my staff, bury it certain fathoms in the earth, and deeper than did ever plummet sound I'll drown my book. [V.i.50–57]

The passage reads something like the one we looked at before, about the globe dissolving and leaving not a rack behind. But now everything looks new, as if even the magic itself had undergone some wonderous sea-change. The dissolving of the globe, the breaking of the staff, and the drowning of the book—they are no longer sad, but turn out to be the true goal of the art. Let us look more closely.

What does it mean for a book to be drowned or for a staff to be broken and buried, for a book to enter into the depths of things but be no longer visible? I believe he means that if we see nobility as having its origins in a play by a man named William Shakespeare rather than in the world itself, then his project indeed would have failed. The difference between the old magic and the new magic is simply in knowing when to disappear.

But how much success does Shakespeare believe his project will have? This is the question faced in the Epilogue:

> Please you, draw near. Now my charms are all o'erthrown, and what strength I have's mine own, which is most faint. Now,'tis true, I must be here confin'd by you, or sent to Naples. Let me not, since I have my dukedom got and pardon'd the deceiver, dwell in this bare island by your spell; but release me from my bands—with the help of your good hands. Gentle breath of yours my sails must fill, or else my project fails, which was to please. Now I want spirits to enforce, art to enchant, and my ending is despair, unless I be reliev'd by prayer, which pierces so that it assaults mercy itself and frees all faults. As you from crimes would pardon'd be, let your indulgence set me free.

What does he mean by the lines

> But release me from my bands with the help of your good hands.
> Gentle breath of yours my sails must fill, or else my project fails,
> which was to please.

Certainly he means to clap, to shout "Bravo!" and to indulge ourselves in the enjoyment of his play. But he may mean more than that. He may mean that if his play does not affect our lives in what we do with our hands, and what we say with our gentle breath for the rest of our lives, then the honor and the nobility will simply go back into their boxes at night, along with the gorgeous palaces and the solemn temples. But—perhaps he does only mean to clap and shout "Bravo!" Perhaps he does indeed think that we become good by enjoying the good.

◆

Shakespeare wrote *The Tempest* and then, true to his word, Shakespeare wrote no more.

Poland

About sixty years ago, a bit more, perhaps, I had a brief conversation with my father. He could remember a time when he was about three and a half and was taken off to visit the home of his grandfather in Biala Podlasca, Poland, some thirty miles from his own home in Brzesc. He spoke of a large house with a small house in back. There was a garden in between, with four pathways leading to a central well. To the right was a pigeon loft. He remembered that well, because it was there they had once been forced to hide when there was the threat of a pogrom. On the other side was an encampment where the cavalry of the Czar's army was stationed. He remembered that too because one of the solders had once put him on a horse and given him a ride.

I returned to Poland once: in some sense I had never been to Poland, and yet this shaggy thing called "I," which lies partly in dead books and partly in grandfather's tales, could return to Poland. I spent the night in Warsaw; and in the morning, together with a friend, a Polish-speaking friend, I took the train to Biala Podlasca. The station was about a half mile from town, and as we walked we spotted a man who seemed to be about the right age. I turned to my friend and said, "He's the right man. Ask him where the Russian cavalry unit was stationed." The conversation lasted about half an hour as we walked. It was all in Polish, and I didn't understand a word, but there were so many hand gestures and nods, that I assumed I had a general understanding of what was being said. I was, of course, totally wrong.

When the conversation was over, my friend told me he had learned that the cavalry camp no longer existed. The neighborhood had been destroyed, and cheap government housing erected on the spot; but still we could go and see the ground on which things had once stood. He also spoke of a Jewish graveyard, and he had some vague information concerning a small museum somewhere in the town.

I went off to the place of the public housing, wandered around a bit, and watched some children at play. But there are always children at play, and nothing meant very much. Then we went to the graveyard. It was an empty field surrounded by a fence.

The dead were gone and their stones were gone. Some say they were used to build the ovens, some say they were used to pave roads. An organization of American Jews who had emigrated from that city some time before or during the war had collected enough money to pay the city to put up a fence and a park. It was a strong wrought-iron fence, but by the time I was there, a part of the fence had been thrown down to make a shortcut for passers-by, and the field still stood empty. There was no park.

We went to the museum. The director said that there had been a mistake. The man I had spoken to was not old enough, and he could only remember where the cavalry had been stationed since 1912; but my story went back to 1904. Again I understood nothing of the conversation, but I followed my guide. This time we seemed to be in an older part of town. The houses were Russian gingerbread, made of boards put together as we once did for log cabins, notched and crossed at the corners. As we turned around, I looked up and saw a street sign with the words "Ulicia Brzesc." I paused for a moment and said to my friend, "I'm lost. My memory is so vague that I don't even know whether I'm actually remembering, or whether this is a story I'm making up on the spot; but I seem to myself vaguely to remember that a distant cousin once told me that there was a strange chiasmal relationship. The memory is that one great-grandfather lived in the city of Brzesc in Biala Russ, while the other great-grandfather lived in the city of Biala Podlasca on a street called Brzesc; and here we are in the city of Biala Podlasca on a street called Brzesc. I think we're here." "Indeed," said my friend, "we are here."

Memories that feel as if they came from the inside are also part of my story. We walked for about two blocks, and suddenly the architecture changed. In the middle of the old wooden houses, there was a large, not very attractive milk factory and one flimsily-built hotel. The compound which housed the factory was very large, and I could not see the end.

"Ah," I said to myself, "that must be the grounds of the old Ninth Cavalry Unit." There was a small yellow gingerbread house next to the hotel. I was terribly disappointed. I know the magnification that takes place in childhood memories, but I had hoped for something more than that. Nevertheless we approached the house, and an old man came out. "Oh, yes," he said, "There is a much larger house on the other side of the compound." We walked and walked, past the hotel and past the milk company, and there it was—the large gingerbread house. We walked around back, and there was the small house and a field of weeds where once the garden must have been. One could still see the four pathways leading to the central circle, but there was no well. It would have been over in the corner by the large house. I later discovered that only one year before, the well had gone dry and had had to be moved. All of this greatly changed my understanding of language. Actions are often preserved in myth and story, but seeing the sounds of an old tale pulling themselves into solid buildings is more profoundly at the heart of the things we have been talking about.

There was a man of about thirty-five in the corner, shoveling coal. He did not look up. We were there for more than an hour. He did not look up. The women and children came out. "No," they said, "no Jews ever lived here."

A camera is the best way making a friend. Few are those who do not want their picture taken, though I sometimes admire them all the more for that. By the door, pictures were taken of laughing people with arms around necks, and I asked the old woman if I might enter—enter what was a front door for her, a crypt ninety years unopened for me. She told me how as a child she had so loved that house, and how she had always wanted to live in it, and now it was hers.

Yes, I too can feel the joy in the old woman's story; it was I who told it to you. And yet I knew that she had known them all, my uncles, my cousins, and perhaps even the old scholar himself, my father's father's father. I also knew that she knew through what events it was that she had come by all that joy.

Memory is a strange beast. As I was working on this passage there was perhaps a half second, no more, in which I reached for a cigarette, although I have not smoked in twenty-five, maybe thirty years. Well, I'm home now, back in Santa Fe.

A Ghost Story[1]

It has often been said in jest, and in derision of a man considered to be a fool, that "he has done nothing more than reinvent the wheel." I hope that during the course of this talk I will be able to adequately articulate why it is that I would give my right eye only to be such a fool; but perhaps none of us is completely up to such a task.

Another way of saying the same thing is to sing that old tune:

> If I could I surely·would
> Stand on the rock where Moses stood.

Now I do not wish to deceive myself into thinking that if only I could stand on the right rock I too would have seen what Moses, or Euclid, or Apollonius saw. I would not have seen; I have not such sharp eyes. On the other hand, it seems to me that to understand anything means to understand how it might have become visible to a human eye. Perhaps more than any other scientist, Einstein seemed to have had a full grasp of the problem. Constantly he was writing papers and articles which dealt in whole or in part with the question of how he came to the notion of Relativity.

According to its title, Einstein's paper on the photoelectric effect is written from a "heuristic" point of view. The paper is about *finding*, in the sense of finding a specific something; and in it Einstein makes clear that seeing anything involves knowing the place from which it can be seen. He writes:

> In what follows, I want to present the thinking and indicate the facts that have led me along the present path in the hope that the point of view associated with these ideas may prove useful to some researchers in their investigations.

For him, answers are wonderful things. They are the things we are all looking for; but he also saw that they have one potentially fatal drawback: Answers tend to erase the question. In so doing they erase themselves as answers, and harden themselves into prejudices and into facts to be taken for granted. On the other hand, it must be pointed out that he published many quite different accounts, and that some of them were even contradictory. For him it was important to see the several different ways in which a thought might have come before a human eye. The question, which of these ways first came to any

1 Lecture given 26 April 2002 at St. John's College, Santa Fe

particular human being, is a question of mere history and is of no interest to him or to us in this evening's lecture.

PART ONE

Plato once recorded a great conversation which is said to have taken place between Socrates and the slave of a by-then-defunct general named Meno. The passage I have in mind begins when Socrates said to Meno:

> Suppose that you call one of your numerous attendants, that I may demonstrate on him.

Meno answered:

> Certainly. Come hither, boy.

We are to imagine a large number of people milling about, some listening, others perhaps not listening; each smiling his own smile or scoffing his own scoff. Plato does not tell us who they were, nor does he describe their several reactions in any of the received manuscripts; but there is a vast labyrinthine tradition from which I believe I have been able to reconstruct, or remember, some fragments of that day. Some say that Euclid was there, but I have my doubts. The timing is all wrong. Euclid lived and thought many generations later. For him the notion of pre-stated postulates was already out there, like a mountain or a blade of wild grass, or like an apple hanging on a tree. How easily that last example came off my tongue while I forgot that no apples worth the eating are given by nature; each of them comes from a grafted tree. Each apple, then, is a product of that original insight into the possibility of grafting, an insight in which I have no share, although I momentarily mistook myself and thought I had all. What I had taken for granted as *just being there* had only been there because of another's labor. And so it may have been for Euclid; and so it surely is for his modem reader. And so perhaps Euclid was not there. At any rate someone was there, and for all I know it was Euclid's grandfather. For the moment, however, let us continue to call our nameless friend by the name Euclid.

Socrates then asked Meno:

> He is Greek, and speaks Greek, does he not?

to which Meno replied:

> Yes, indeed; he was born in the house.

Euclid was pleased. "If they both speak Greek," he said to himself, "they will not only be able to speak about geometry, they will be able to speak *geometry itself*. The boy, of course, will have to learn many strange new words, words not found on the everyday tongue; but they will be Greek, and so he's off to a good beginning."

◆

"No," thought Descartes, "no, the fact that he can speak Greek just doesn't sound all that important to me. Ever since that fateful afternoon when I first saw that I had to take more seriously, and make the center of my concern, that curious relationship (which I had originally no more than noticed) between algebra and Pappus's concept of a *locus*, human thought and its relation to what it perceives has taken on a wholly new cast. If the human mind sees itself as radically other than the world it perceives or imagines about itself, and forces itself to think entirely within its own terms, it is capable of a kind of clarity and distinctness of which no man has ever dreamed. A *circle* with the fullness of its roundness implies divinity to some, and simple boredom to others; but the clarity of the *equation* is the same to all. Mathematics cannot be spoken in Greek or Latin or in French or in German. Mathematics can only be written in mathematics itself." And with that, off he went to sit by his stove and work things out, paying no attention to the fact that the whole of his new enterprise rested on a decision he had made in normal everyday French.

◆

Leibniz gave an obliging smile and thought: "Descartes should never have walked away. He's right, of course, mathematics must be written in mathematics; but it must also be spoken: in Greek and in Latin and in French; and in German, yes, and in Chinese, too. Those who demand clarity understand that each language—and in fact each person, precisely because they are who or what they are—will have a slightly different view even of what *addition* is. They say: "one *plus* one *is* two" or "one *and* one *are* two," or even "one *and* one *make* two," and sometimes even "one *and* one *makes* two." Each of these expresses a slightly different understanding of what is involved in the act of addition. That is why they prefer the equation. However, Descartes' clarity comes at the price of narrowness. It fails to see that each of those views reveals only one facet of the whole, a whole which no man may ever see with Cartesian clarity."

◆

Socrates and the boy continued their conversation:

> Soc: Tell me, boy; do you know that a figure like this is a square?
> Boy: I do.
> Soc: And you know that a square figure has these four lines equal?
> Boy: Certainly.
> Soc: And these lines which I have drawn through the middle of the square are also equal?
> Boy: Yes.

Euclid then burst in: "But Socrates, the boy doesn't really know any of those things. Why, the construction of the square alone requires five postulates, five common notions, eighteen propositions,[2] and God knows how many definitions!"

"No, my dear friend," said Socrates, "the boy knows many things. He knows, for instance, that if two sides of a triangle are equal the angles at the bottom are equal as well. He also knows that that truth has nothing to do with those silly little lines you've drawn underneath, but comes from something he fully grasps and to which those more sophisticated than he give the name 'symmetry.' These things he knows as well as you know your five postulates; how well that is, we shall not at present ask.

"He would also know that if there are two sets of lines laid alongside each other as in Figure 1, then AB equals CD and AC equals BD; but he might not know at first that ABCD equals ABFE. However, if the lines AC, CE, and EA should happen to fall together in his eye as a triangle, then he might see that the triangle ACE is equal to the triangle BDF because it's the same thing just slid over. And if he were curious enough and had the kind of eye that could hold both the triangles and the quadrilaterals in mind at the same time, he would be in for quite a thrill.

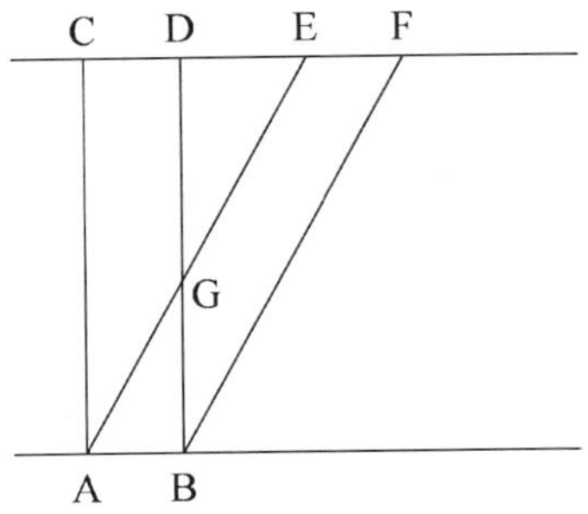

Figure 1

"And so you see, my dear man, there is a whole world of, let us say, five hundred things that the boy already knows and from which he will go on to discover much more. I pick the number *five hundred* partly at random and partly as a private joke."

◆

Euclid was not much impressed by Socrates' speech and thought that no science could be founded on such beginnings. But for the moment at least, he held his peace.

Then Socrates and the boy went on to rediscover the double square. But did the boy need Socrates? Socrates was not always available. There must have been someone who found it on his own. The question then arises: "Can I find a spot from which the double square might have become visible to a human eye, or must I too forever rely upon Socrates?

In one sense the solution to our problem is not so far off. Who has not stared at a tile floor, as in Figure 2, watching the patterns dance as the lines

2 Propositions 1, 2, 3, 4, 5, 7, 8, 11, 13, 15, 16, 23, 26, 27, 29, 31, 34, 46

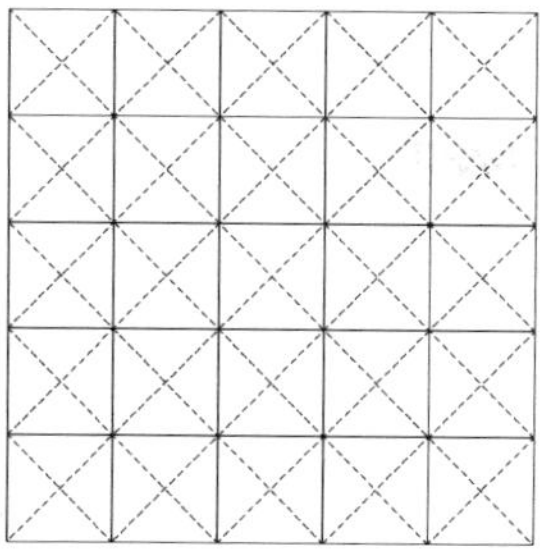

Figure 2

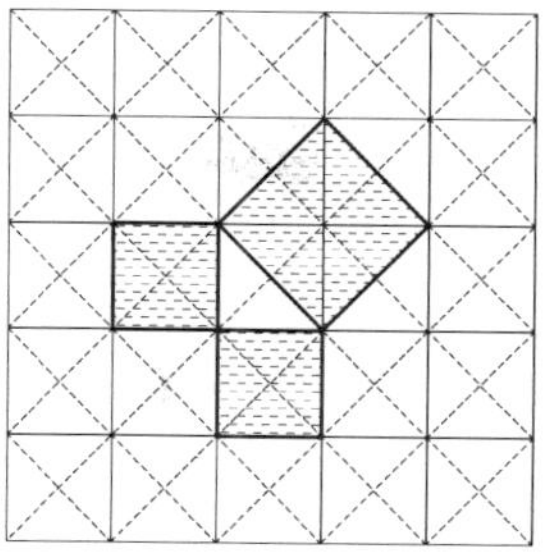

Figure 3

form themselves first into one group and then into another? Some may even have seen it as we have marked it in Figure 3.

Surely many eyes saw that square; but it must have taken an extraordinary man to see that it was something worth remembering and being concerned about, even worthy of telling one's neighbor. Perhaps that was because in some way he realized that when he said the lines were equal, he meant something quite different from what Homer meant when he wrote the line:

> When they had done all this and had prepared the feast, they ate it;
> and every man had his full and equal share.

This equality felt different because it did not depend upon the skill of the stonemason or tile-setter but merely upon the pattern or *eidos* that he had intended. For that reason it would always be there, even for the neighbor, even for all generations to come.

Here is another thing that calls forth wonder. There are those rare individuals who, whenever they see something, ask themselves: "Is this a simple truth, or is it a minor case of something larger, stranger and even more wonderful than itself?" Some such a one must have asked himself, "What if the square had no longer been a square, but a rectangle instead?" He redrew the figure as best he could, and so discovered the square on the diagonal to be equal, not to *twice the rectangle* (Figure 4), but rather to *the sum of the squares on its two sides*—since each in its own way was equal to the large square minus four times the triangle (Figure 5).

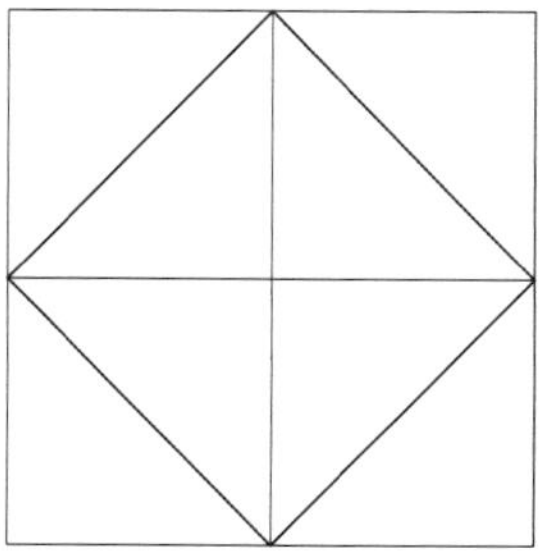

Figure 4

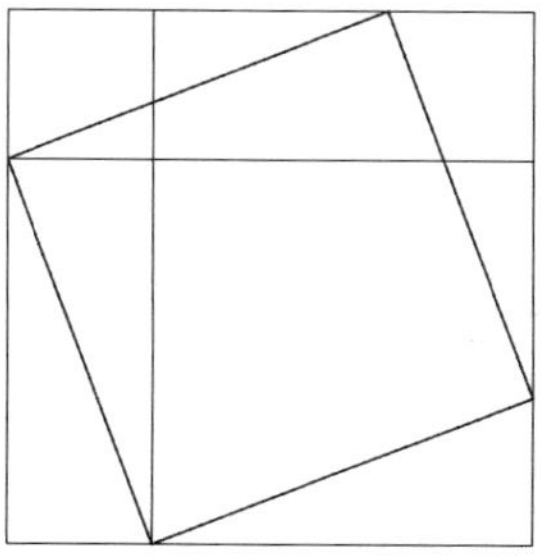

Figure 5

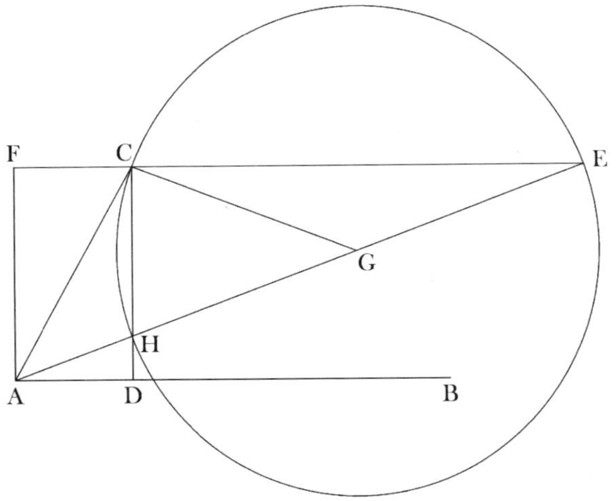

Figure 6

Now we must go back and ask ourselves why Euclid was so distraught. What did Euclid (or whoever it was) see that caused him to abandon that world which Socrates and the·slave boy held in common so comfortably, and to replace it with five postulates all laid out and agreed upon beforehand? Of what was he so afraid? Perhaps it was something like this:

Almost since the beginning of time, carpenters, and hence geometricians, have known how to bisect an angle. Then, according to Pappus, one of the ancients proposed the following construction for trisecting an angle:

> Given angle CAB (Figure 6), it is required to trisect it.
>
> Let CD be perpendicular to AB; complete the parallelogram CDAF, and extend FC indefinitely .
>
> Let AE be drawn to intersect CD at H and FC extended at E, such that HE shall equal twice AC; and let G be the midpoint of HE.
>
> Now since $\angle HCE$ is a right angle, G is the center of the circle passing through points H, C, and E. Thus $GH = GC = GE$
>
> Therefore $\angle EAB = \angle CEA = \angle ECG$
>
> And $AC = CG$, so that $\angle CAG = \angle CGA$
>
> But $\angle CGA = 2\angle CEG = 2\angle CEA = 2\angle EAB$
>
> Therefore also $\angle CAG = 2\angle EAB$,
>
> that is, $\angle EAB = 1/3\ \angle CAB$
>
> Q.E.F.

Could you all follow the proof? Do you then all agree that I have succeeded in trisecting the angle ABC? No? Well, I guess we'll have to take a vote. Will those who accept the proof indicate by saying *aye*? Opposed? The problem, of course, is whether you are inclined to allow me to draw the line AHE so that

HE is double AC. Many were so inclined; and, to be truthful, I do not know where I would stand had Euclid not been our tutor.

But what does it mean to have *voted* on the validity of a mathematical proposition? Have we not, by doing so, just destroyed even the possibility of any meaningful mathematics? Now I am not prepared to read Euclid's reconstruction of geometry—artful and hence unnatural as it may be—unless I am prepared to see the crumbling away of mathematics in its entirety. This will require replacing our common world, with its miscellaneous hundreds of striking and inviting but not obviously interconnected relations, with a world based on *five* pre-stated postulates laid down by one man and agreed to by all. It may also mean starting with the kind of *definition* that Meno was looking for, and in which Socrates saw so much that was doubtful. There may, however, be no other way; and the beauty of what has been built on this edifice is surely undeniable.

Euclid himself, it must be pointed out, was not wholly committed to these new beginnings. For instance, he still assumes, without a postulate, that if a straight line contains at least one point lying *within* a circle and at least one point lying *outside* the circle, then it also contains at least one point lying *on* the circle.

Was it precisely the *trisection of the angle* that moved our Euclid, whoever he was, to found geometry upon a limited number of pre-stated axioms and postulates? Possibly not; it could equally have been the discovery of irrationality. But so far as I can see, only some experience such as we have just gone through could have given rise to the issue. As far as the exact proof is concerned, not only is it unclear that such merely historical knowledge is available, its value in most cases is even dubious. Far more important is our realization that *our mode of thought itself* is the hard-earned fruit of human thought, and not a thing to be taken for granted as given.

◆

The Meno began when Meno asked the question:

> Can you tell me, Socrates, whether virtue is acquired by teaching or by practice; or if neither by teaching nor practice, then whether it comes to man by nature, or in what other way?

Most of our thoughts, such as the need for pre-stated postulates, represent all the alternatives Meno listed. They are like ghosts, ghosts of the thoughts and concerns of other men, which flit from mind to mind, taking up residence in houses within which they were not born, and masquerading as living beings. Homegrown thoughts come to be within that particular but fuzzy horizon we call our mind. They trail off into the periphery of our horizon; and, if pushed, can find those chinks in the wall which lead beyond any horizon, back to those foggy primal inklings that first gave rise to our full-blown thought. The ghosts of other men's thoughts reach out for their own craggy horizon; and, not finding it, can only return to the center, where they congeal into a hard rock of sedimentation. Many magnificent and even beautiful structures can be built on that rock; witness modem science. Those structures are no less the products of genuine insight than were their predecessors; but they rest upon the ghost that has become a rock—a living thought in the teacher, but in the student solid and unquestionable.

This is the world which Descartes wished to set aside by a simple act of universal doubt. But the ghosts bury themselves too deeply into our souls to be whisked away in the twinkling of an eye. Ghosts can be exorcised only by bringing them back to life as *living thoughts being thought through* as objects of serious concern.

◆

At the end of their conversation, when most of us were firmly convinced that the boy had finally understood how to find the double square, Socrates surprised us by saying:

> At present these notions have just been stirred up in him as in a dream; but if he were frequently asked the same questions, in different forms, he would at last know as well as any one.

Euclid could no longer contain himself and finally spoke up. "Now, Socrates, you see where all of this has led. When *I* prove a proposition, by God, it *stays* proved!"

The two spoke no more on that day; but as they parted, Socrates was thinking about what Leibniz had said, while Euclid was concerned about the man who thought he had trisected an angle.

PART TWO

The ancients regarded Apollonius's *Conics* as one of his minor works; and I must say that I tend to agree with them, although I consider the *Conics* to be one of the two or three most insightful book of mathematics I have ever read. They thought of Apollonius primarily as an astronomer, whose great work was his now-lost book on astronomy. To understand this assessment we must take a brief look at the state of astronomy before Apollonius came along.

As you know, the Babylonian priests had diligently kept daily records of the motion of the planets over a period of many centuries. To me it is incomprehensible that men should have devoted themselves to such a task when no discernible pattern could possibly have emerged for generations! What prompted them? What kept them going? Perhaps it was a conviction that wherever a hint of order shines forth, more will appear if one is sufficiently dogged in one's looking.

In the case of the wanderer Venus they found that, while the star more or less kept pace with the sun, it was sometimes to the north, sometimes to the south. First it would run ahead of the sun, that is, to the east; then it would slow down till it began to lag behind the sun, to the west. Then it would speed up and start the whole thing over again.

The only problem was that the distances and times of these sallies kept changing, with no apparent order to them. Now, you have to remember, these priests were men who had no tradition of science behind them; nevertheless they continued looking and recording until they were satisfied that that highly erratic series began to repeat itself.

While the Babylonians, so far as I can see, did have some notion of the stars as eyes[3] which were set in wheels that turned because the spirit was in the wheels, they had nothing that could address the back-and-forth motion of Venus. This was left for a Greek named Eudoxus (Figure 7).

He set the axis A into a sphere that moved with the sun, so that if nothing more were done, the star (at point O) would move with the sun. However, he added two more motions. One moved clockwise around the axis A, while the other moved at the same speed around the axis B, in the contrary direction. As you can see, if the two poles were set along the same straight line, the two motions would cancel each other out and the star would remain at the point O. If, however, the poles are inclined to one another, the motion towards D will pull the star somewhat to the south, but now the motion to the east will no longer overcome the motion to the west; and the star will end up at point I. It will then go on to make a quasi figure-eight. As you can see, the star will sometimes be above the sun, sometimes below. At times it will be ahead of the sun, at times behind. However, according to this construction, the star would have to complete exactly two back-and forth motions for each

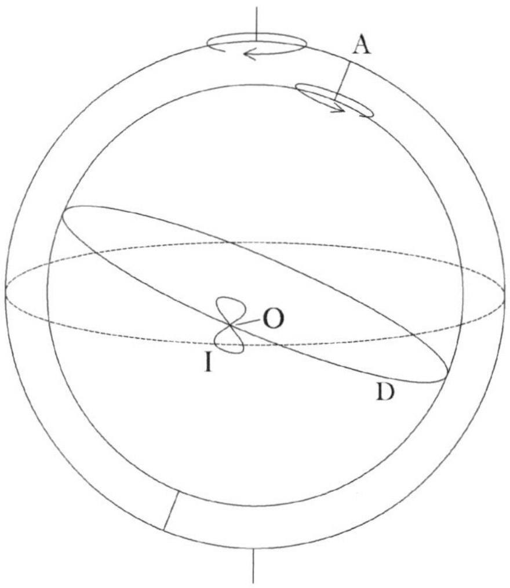

Figure 7

3 Cf. Ezekiel 1:17–21

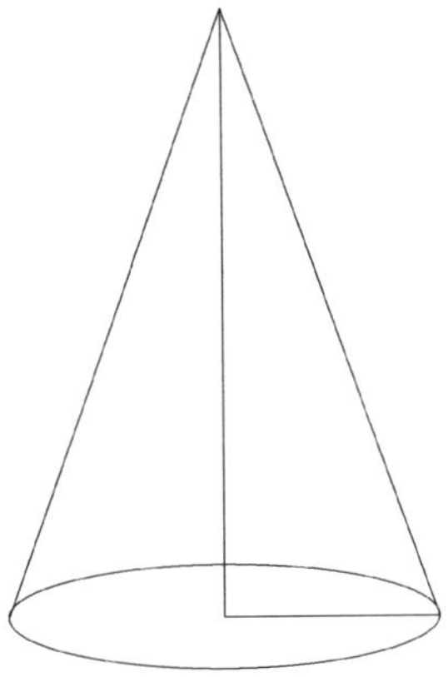

Figure 8

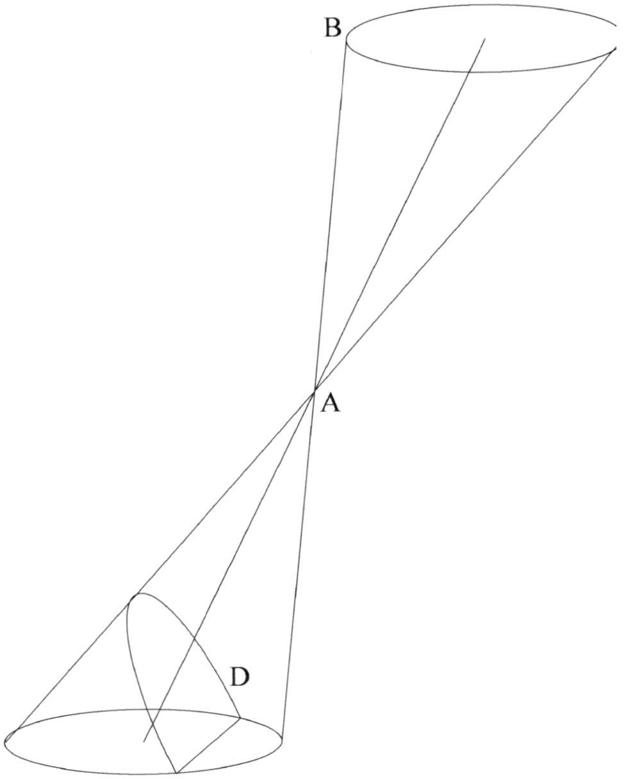

Figure 9

up-and-down motion—which is very far from the truth. Eudoxus succeeded only in accounting for the kind of thing that happens, and made no pretense whatsoever to accuracy.

Another way of saying this is that for Eudoxus, the *phenomena*—the things to be accounted for—were the *phenomenal* things, wondrous things such as retrogradation, eclipses, and the like; whereas for Ptolemy the phenomena were the *ordinary* things. That, for Ptolemy, meant accounting for the appearance of each day according to its day. If that work were done well, the *phenomenal* things would take care of themselves. How this change might have taken place, we have yet to see.

With that in mind, we must next turn to the *Conics* of Apollonius.

◆

Prior to Apollonius, the cone was thought to be generated by revolving a *right triangle* about one of its legs (Figure 8). Apollonius, on the contrary, defined it as the surface generated when a *line* such as AC (Figure 9) is rotated in such a way that the point A remains constant while the point C moves along the circle CB.

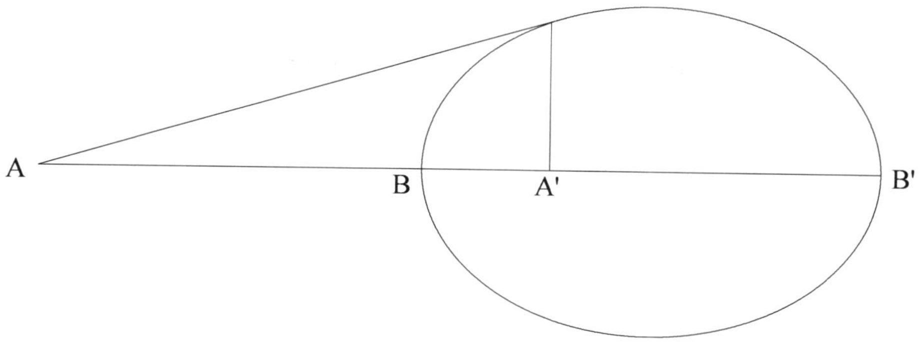

Figure 10

I believe, by the way, that what originally sparked his interest was his need, as they say, to *build a better sundial.* The circle BC was the daily path of the sun, A was the tip of the pointer, the cutting plane was the plate of the sundial, and the conic section was the path of the shadow to be calculated for some given day. But all this was soon forgotten, and we can see the act of thought at work when we see how fascination replaces need.

In Book One, Proposition 36, Apollonius proves that if from a point on a conic section a tangent and an ordinate be both drawn to the axis, then in Figure 10,

$$AB : AB' :: A'B : A'B',$$

that is, the point A cuts the line BB′ on the outside in the same proportion that A′ cuts it in the inside.

For many reasons this ratio is called the *harmonic ratio*. For us, the harmony lies in the fact that if we consider the line BB′ and take the point A′ on it, if A′ is, say, seven times closer to B than it is to B′, then A is also seven times closer to B than it is to B′. In that sense, the two points are *in harmony*. It should be clear that as A′ is taken ever closer to the center, AB can become as long as you please.

The first question that always enters my mind when I read something like Proposition 36 is: "How did he smell out that such a thing even might be true? Where was he standing when such a possibility occurred to him?" But of course that is merely a folksy way of putting the question; and so stated, it is of no more than historical interest. The real question is: "How might such a possibility have become visible to the human mind *as such*?" Of course, "to become visible" means to become visible from within a *specific*, and hence a *historically specific*, horizon. In what sense this is true, I hope to make somewhat clearer in the course of this talk.

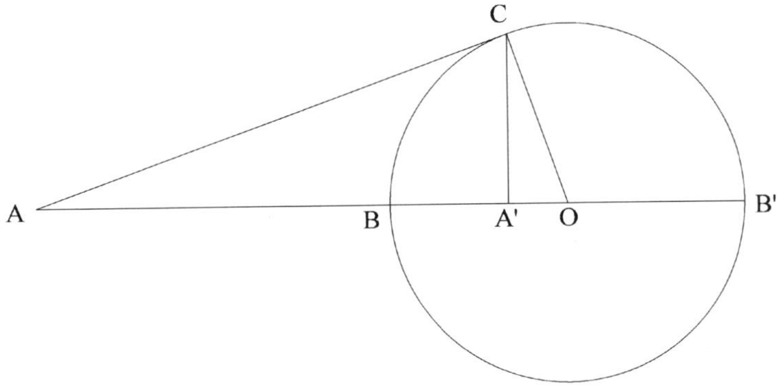

Figure 11

In the special case of the *circle* (Figure 11), it is pretty clear that

since $$\triangle AOC \sim \triangle A'OC,$$

$$AO : CO :: CO : A'O$$

or $$AO : BO :: BO : A'O$$

that is, $$AO, A'O = BO^2,$$

which is what Apollonius proves for all the conic sections in the following proposition, Proposition 37. The harmonic ratio for circles then easily follows from this relation. For from the above,

$$AO : OB :: OB : OA' \quad \text{therefore} \quad AO : OB' :: OB' : OA'$$

Then, componendo,

$$AB : OB :: A'B : OA' \quad \text{and} \quad AB' : OB' :: A'B' : OA'$$

Alternating,

$$AB : A'B :: OB : OA' \quad \text{and} \quad AB' : A'B' :: OB' : OA'$$

But $OB' = OB$; therefore

$$AB : A'B :: AB' : A'B'. \quad \text{Q.E.D.}$$

All this about the harmonic ratio, of course, is still based on the hint we·took from Apollonius. We still do not know what attracted him to this proposition. What made him think that it might hold for all conics in spite of the fact that $\triangle AOC$ will no longer be similar to $\triangle A'OC$? The fact that point C no longer plays any role in the final statement led me to construct a number of ellipses on the same major axis, as in Figure 12.

This is what I've come to call the *nutcracker view* of the ellipse. That, I thought, was a lot of fun and might be useful one day—but it just didn't quite smell like a *beginning*, something a keen eye could have picked up just because it was there. As I continued to gaze at the figure, however, it suddenly transformed itself into a *sphere resting in a cone*. That was enough to *guarantee* that

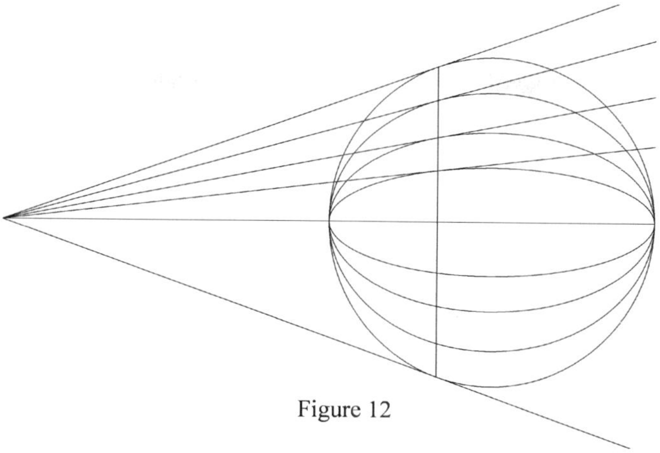

Figure 12

the tangents to the ellipses would all meet the diameter at the same point. That, it seemed to me, was something an eye keener than my own might have picked up; and I saw that a mind thirstier than my own might have been sufficiently struck by what it had seen to search out that ratio by using the case of the circle. All this seemed even more likely to me when I remembered that Apollonius was an astronomer and, as such, would have spent late hours drawing many such figures; so he would have known that spheres are drawn in perspective by drawing ellipses, as was done in Figure 7.

A mind like that of Apollonius is the true meeting of Dionysus and Apollo. Look at any one of his propositions, and imagine the wildness of his thoughts as he searched for what might be true. Then consider the sobriety with which he labored to re-establish those truths on good Euclidean grounds.

Let's look back for a moment to see where we are. We have a Euclidean proof that, in a circle (Figure 11),

$$AO, A'O = BO^2,$$

from which we derived the harmonic ratio. We also know with pre-Euclidean certainty that these things are also true of the ellipse; but we have no Euclidean proof. We are also sure that if we ever find such a proof, the hyperbola will either come along quietly, or else something equally interesting will happen.

We must now begin to look for such a proof. But before doing so, let us compare the normal Euclidean proof with the normal Apollonian proof. Most of Euclid's longer propositions refer back to many earlier ones, whereas Apollonius rarely refers back to more than one earlier proposition, except for those that deal with the handling and transformation of ratios. To that extent, the art of Apollonius is different from the art of Euclid. Hence, his art of searching may also differ from that of Euclid.

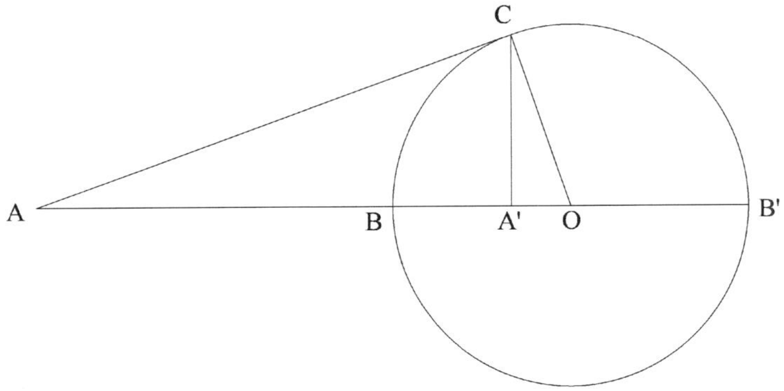

Figure 11, repeated

The first thing he must have noticed is that the proposition, so easily proved for the circle, cannot be proved quite so easily in the case of the ellipse. As it happened, he was forced to turn the whole thing around by establishing the harmonic ratio on its own, then deriving Proposition 37 from it. This longer route, he found, required the whole of Book One up to that point. In fact, it might be wise to view Book One precisely as the product of his search for that proof; I sometimes call it "rummaging through Apollonius's wastebasket."

And so at this point I shall leave you to follow his backwards journey to see how he managed to arrive at his beginning. Before going on, however, let us take one more look at the figure of Proposition I.36 for the case of the circle (Figure 11), as Apollonius himself might have done.

Memory is strange and awesome. Augustine speaks of

> the plains and caves and caverns of my memory, innumerable
> and innumerably full of innumerable kinds of things.

He thinks of it as a place where

> images are with an admirable swiftness caught up,
> and stored as it were in wondrous cabinets.

But the conceit of memory as a place or mansion where things are kept on shelves or in bins or on disks—tokens like pressed roses, or little slips of paper waiting to be picked up by a passing hand—never seemed right to me. To me, memory has always seemed more like an activity than a thing; more akin to a *low-level thought* than to the pressed rose itself. At some level we must always be thinking every thought we have ever thought, seeing everything we have ever seen, hearing everything ever heard. Memories often seem more to come to us than we to them. We can almost sense such activity going on when we say "The word is on the tip of my tongue."

Apollonius, as we supposed, first became attracted to the study of conic sections through his interest as an astronomer in building a more accurate, more intelligible sundial. But from what Ptolemy says of him, I think something quite different began to happen as he took still another look at the diagram, our Figure 11.

Those old memories of Eudoxus began to stir more rapidly. But had they not been alive in him at some low level, they could not have come looking for him nor could he have gone looking for them. If point A were the earth, and if the point O were to move evenly with the mean position of the sun, then the planet Venus would be free to travel at its own speed around the circle BB′. In this way one could "tune" the motion to achieve a kind of accuracy Eudoxus never dreamed of. From what Ptolemy says of Apollonius' general interest in the relation of inside to outside, as well as what we can see in the *Conics* itself, it is clear that Apollonius knew that the earth could be understood to be at the point A′ as well, especially in the case of the outer planets. And if he knew that much, God knows what else he knew.

Normally we think of men as having questions, for which they then go on to look for answers. I should like to suggest that almost the opposite is true. Questions only arise when we *almost* have an answer. Neither order nor chaos can give rise to a question. As it was for Job, so it was for Apollonius, and so it is for all men. There are things that make too much sense to make no sense, and yet they make no sense. These are the things that give rise to questions. In the evening and in the morning is wonder born.

PART THREE

In Proposition Sixteen of his *Theory of Parallels* Lobachevski writes (Figure 13): "In the uncertainty whether the perpendicular AE is the only line which does not meet DC we will assume it may be possible that there still other lines, for example AG, that do not meet DC how far soever they may be extended."

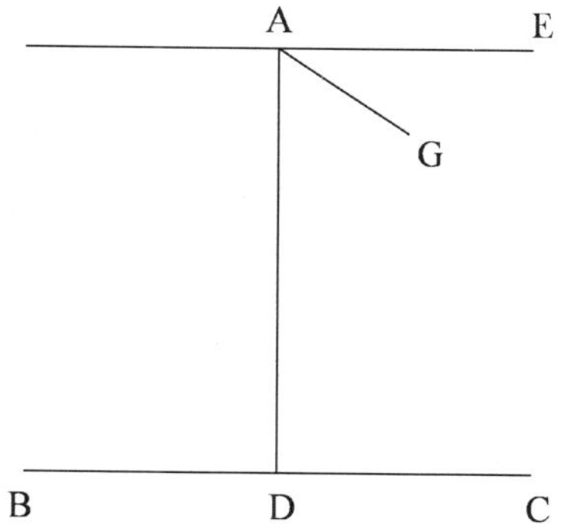

Figure 13

We read, but we do not understand—unless we, like the author, have already felt that uncertainty for ourselves. We could read on; and if we did, we would find ourselves in the middle of a strange new world. Soon, perhaps all too soon, we would begin to be able to navigate the seas of this new world and even discover some of its as yet uncharted islands. But there would be a gap in our memory. It would be as if we had .arrived in the new world by a magic carpet and never travelled the road from home on foot. We would never have seen the breach or chink in the horizon of our old world which caused our fathers to go beyond them.

I remember that when I was a Junior, someone—it may have been Jascha Klein, related to me a saying which has been attributed, at one time or another, to almost everyone who lived in the Seventeenth Century. It says: "The ancients were giants, and we are but midgets; but since the midgets stand on the shoulders of the giants, they can see farther than did the giants." I felt that it was very true, but that those who were most fond of saying it did not understand its fullest consequences. There was one thing which the giant saw, and which the midget could not see: the midget could not see the ground upon which he himself rested, because the giant's head was in the way.

Our path begins with a remark made by a man named Proclus; but to understand it we must remind ourselves of some things in Euclid.

Proposition 27 of Book One reads:

> If a straight line falling on two straight lines make the alternate angles equal to one another, the straight lines will be parallel to one another.

Proposition 29 of Book One reads, in part:

> A straight line falling on parallel straight lines makes the alternate angles equal to one another.

As you no doubt remember, Proposition 29 relies on the fifth postulate, whereas Proposition 27 does not. I remind you (but only parenthetically for the moment) that Proposition 27 does, however, rely on Proposition 16.

Proclus was outraged. In his commentary to Proposition 29 he grumbles:

> Not everybody admits that this generally accepted proposition is indemonstrable. For how could it be so when its converse is recorded among the theorems as something demonstrable?

Proclus had already said of the fifth postulate:

> This ought to be struck from the postulates altogether. For it is a theorem ... and the converse of it was proved by Euclid himself as a theorem.

For Proclus there was a fundamental asymmetry in the foundations of geometry, which, if not corrected, would soon cause its downfall. To avoid this downfall it was absolutely necessary to come up with a proof of the fifth postulate.[4] To understand ourselves and where we stand, we must follow that pursuit.

4 Proclus was right, but only to a certain extent. There was a certain asymmetry in Euclid, and it did lead to his downfall from the unique position he once held. However, the asymmetry was not the one Proclus had in mind.

As we pointed out, Proposition 27 relies on Proposition 16, which presupposes that two straight lines cannot have more than one point in common. It is easy to see that this is in fact

A thousand years went by, during which time countless people tried their hand at proving The fifth postulate, mainly by *reductio ad absurdum*. Many strange and absurd things were indeed proved; but each time it was shown that the results of denying the fifth postulate only contradicted a theorem in Euclid which had *itself* been established by means of the fifth postulate. Strictly speaking, none of these admittedly bizarre propositions could be disproved on the basis of the first four postulates alone. Nothing had been gained. To put it metaphorically, it was as if many chimerical houses had been built with the intention that they fall—but none of them fell. Nevertheless, no one ever supposed any of them to be other then fantastical.

But in 1733 Girolamo Saccheri published the book EUCLID AB OMNI NÆVO VINDICATUS, or *Euclid Freed of Every Fleck*. Unlike those who came before him, Saccheri was fully aware that Proposition 27 rested on the hidden postulate. He was therefore no longer troubled by the supposed problem of asymmetry. He began, in effect, by replacing the fifth postulate with two hypotheses. The first, which he calls the *obtuse angle hypothesis*, states:

> Through a point not on a given line there cannot be drawn any straight lines which do not cut the given line.

The second, the *acute angle hypothesis*, states:

> Through a point not on a given line there can be drawn more than one straight line which does not cut the given line.

Ostensibly, he is out to prove that both the obtuse angle hypothesis and the acute angle hypothesis are, as he says, "absolutely false."

---

the counterpart of the fifth postulate. If we look again at Figure 13 and ask ourselves how many straight lines can be drawn through the point A which do not cut BC, we might begin by answering "either one, or many, or none." The notion that there could be *many* is negated by the explicitly stated fifth postulate; but let us examine the notion that there might be *none*.

Consider, then, the perpendicular AE. Will it meet BC on the right or on the left? Symmetry would say that neither by itself can be true; and so, if it must meet, then it must meet on both sides. The true asymmetry lies in the fact that there are two symmetrical postulates: one explicitly stated (the fifth postulate), and the other, that two lines can meet *in at most one point*, left unstated. But, I repeat, Proclus seems not to have noticed that; why did Euclid? From something Proclus says, I suspect that the notion of *lines that seemed to converge but did not meet* was new and paradoxical, and that Euclid found it necessary to postulate that no such fate was in store for straight lines.

In other words, the asymmetry Proclus felt lay not in a need to prove the fifth postulate, but rather in the need for a sixth postulate which would read: "No two straight lines can meet in more than one point." The proposed postulate would have a more secure standing within the human imagination: it would seem to require no more reflection than does the notion that if one point on a straight line lies within a given circle and another point lies outside the circle then the circle and the line must have at least one point in common.

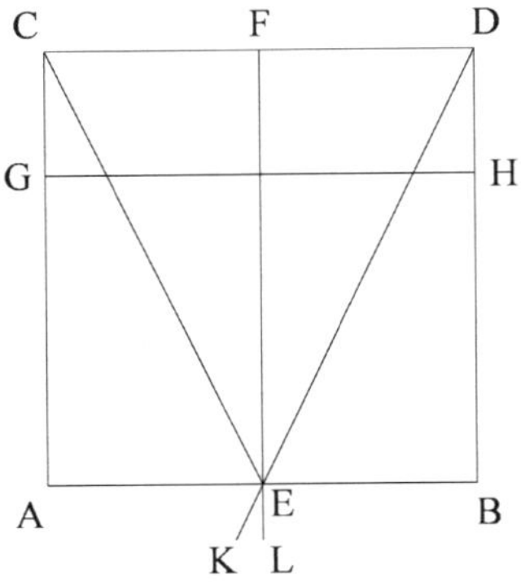

Figure 14

In his fundamental figure (Figure 14), AC and BD are taken equal to each other and at right angles to AB, with E and F the midpoints of AB and CD. First he proves

$$\angle ACD = \angle BDC$$

and that the angles at E and F are right. Next he proves that

IF	THEN
$\angle ACD < = >$ a right angle	$CD > = < AB$
$CD > = < AB$	$\angle ACD < = >$ a right angle
If either hypothesis is true in one case	it is true in all cases
$\angle ACD < = >$ a right angle	$\angle KEL > = < \angle EDB$

Then in Proposition Fourteen he proves:

> The hypothesis of the obtuse angle is absolutely false, because it destroys itself.

However, he explicitly uses Euclid 1:17, which he knows to contain the hidden postulate and which therefore makes it useless for proving the proposition. Saccheri's very next proposition, Proposition 15, reads:

> By any triangle, of which the three angles are equal, greater, or less than two right angles, is established respectively the hypothesis of the right,[5] obtuse, or acute angle.

As we suspected, the hypothesis of the obtuse angle is as alive and well as ever, and the same fate awaits the hypothesis of the acute angle later in the book.

I now believe I understand Saccheri's rather strange title. Does one "vindicate Euclid of every fleck" by proving the fifth postulate, thus showing him to be a fool for having postulated it? No, rather, by showing that Euclid was right in forging such a postulate, because it could not in fact be proved!

5 Equivalent to Euclid's fifth postulate.

Unlike those before him, Saccheri had a global understanding of each hypothesis. The fantastical houses had pulled themselves into two distinct, though fantastical, cities. The unity of each city lay in the fact that he could guess beforehand which building would stand in which city. It was this unity that made the fantastical less fantastic, even thinkable. However, he had no proof of consistency, but only a pre-Euclidean sense of their unity.

Lobachevski, by establishing a trigonometry, proved that the new geometry is consistent if our number system is consistent. However, the consistency of our number system still rests on our pre-Euclidean sense of unity, and Gödel's Incompleteness Theorem seems to show that it always will.

And so it is that no one ever went out in quest of a promised land unless he had already had an inadvertent glimpse of it from high upon a mountain top.

Ptolemy as Teacher[1]

It is often difficult to be sure what someone means when he speaks or writes to us. The words, "There will be a full moon tonight," mean one thing from the mouth of a lover, but something else in a local radio weather report. One must know something about the soul of the speaker to grasp his words, and yet many times it happens that our only path to the soul is through those very words themselves. In such a case we can only compare or contrast what is said in one place with what is said or left unsaid in another, till we reach some understanding of the author, or at least reach some clarity about the subject of our own confusions and the alternatives open to us. I believe that this kind of problem may appear in the works of Ptolemy, and in this article I shall try to say why that is so.

Ptolemy's *Syntaxis*—now more familiarly known to us as the *Almagest*—is a book addressed to a private man, Syrus. Ptolemy begins by telling him something about the "trueborn philosophers." We are not told whether Syrus belonged to that noble race or not, but Ptolemy's wording would seem to imply that the addressee is of a humbler nature. Although we shall not have much to say about the Copernican Revolution in the course of this article, such an investigation might well begin by contrasting Ptolemy's Preface to the Almagest with Copernicus' statement that "mathematics is for mathematicians."

According to Ptolemy, the trueborn philosopher separates the theoretical part of philosophy from the practical—that is, the part which purely desires to look and to understand from the part which deals with human actions: ethics and politics. To this traditional distinction, Ptolemy makes a small addition: "In spite of the fact that *praxis,* the way men act, seems somehow to be dependent on *theoria,* the way in which men look at the world, they differ in two important respects. Knowing how to act well is within the proper domain of all mankind, whereas *theoria,* which lies behind *praxis,* comes only with the time, insight and patience that study requires." The other distinction is that the theoretical sciences are progressive, whereas *praxis* is "fostered by repetition." Ptolemy seems to mean this in a double sense. When we study Euclid, we begin with the axioms and go on till we have mastered Book 13, then we begin more advanced work; but in ethics, we grow by meeting the same problems over and over again.

1 Originally published in *Au Verso* no. 6, St. John's College, Santa Fe (1975).

At the end of the Preface,[2] and again in many places throughout the *Almagest,* Ptolemy reminds us that any improvement he has been able to make in the science is due to diligent application on the part of others who came before him. Ptolemy can account for the way things appear better than Hipparchus can, precisely because Ptolemy has Hipparchus' work to build upon.[3] In making the distinction between the progressive and the repetitive sciences, Ptolemy perhaps means that, with regard to the theoretical, each generation begins on a higher plane than their fathers began on, whereas each child must learn to be good all over again from the beginning. Ptolemy goes on to draw the following conclusions in his reflexions:

> Therefore we have judged it more fitting to arrange our deeds [*praxis*] so as never to lose sight, even in the smallest thing, of that which would contribute to the beauty and the order of our inquiry through the application of our imagination; and for the most part to use our speculation [*theoria*] for the exposition of numerous beautiful theorems, and particularly those which are called mathematical.

According to Ptolemy, the distinction between the two kinds of science partly lay behind his decision to act in a certain way, that is, to write a certain kind of book—a book which would present well-ordered theorems arrived at through *theoria.*

Ptolemy then divides the theoretical sciences into three groups: the physical, the mathematical, and the theological. In what may ultimately appear to be a rather ironical way, Ptolemy chooses a path which superficially reminds one of Descartes' path, the way of mathematics. Both physics and theology have their relative advantages and disadvantages. Physics deals with that aspect of the world which is most immediate to our senses, but that aspect is also one which does not lend itself to being grasped with any precision or clarity. Theology, while it may deal with the highest things, is according to Ptolemy "in no way phenomenal or attainable." Only the mathematical, both from the standpoint of the thing itself and from the point of view of man, can be grasped with intelligibility. And only it can lead man to the highest things, to nobility of action. Ptolemy completes the Preface by stressing the importance of the work of his predecessors and playing down his own achievements. The *Almagest* intentionally lacks any revolutionary character. It is merely the intimate working out of what has already gone before.

As Ptolemy presents his arguments concerning the spherical motions of the heavens, it becomes clear that he had spent a good deal of time reflecting

2 Ptolemy's Preface is called Chapter 1 in the Encyclopædia Britannica edition of the Almagest.

3 Book VII, Chap. 1

upon the way in which the world might first have appeared to the most primitive of men. In the beginning there seemed to be two radically different kinds of stars: those which are always visible and those which seem to arise out of the earth in the east only to disappear later in the west. Ptolemy indicates that it is the quasi-regularity of the second kind of star which first gave men the impetus to keep records of the stars, and which is the root of the drive towards precision within the science. One might even say that these stars lie behind the Babylonian endeavors in astronomy. Those other stars, the ones which go in perfect circles and are always visible, seem to be more Greek in character, since they would more reasonably lead to a geometrical account of astronomy. After this reflection, Ptolemy adds a curious statement:

> And then they saw that those nearest the always-visible stars disappeared for a short time, and those further away for a longer time proportionately, and for these reasons alone it was sufficient for them to assume this motion [the spherical movement of the heavens] as a principle; and forthwith to think through all the other things consequent upon these same appearances in accordance with the development of the science.

Ptolemy here seems to indicate that the man who first noticed that the stars on his northernmost horizon set only for a short time, and so could be understood as being a kind of continuation of the always-visible stars—provided that there was an "underneath" to the world—was the one who began the science of astronomy as such. For Ptolemy, the discovery that stellar motion could be rendered intelligible by conceiving it to be continuous was the only true moment of human insight; the later addition of epicycles and eccentrics was a mere carrying through of that original insight. One might even say that since astronomy is the oldest of the exact sciences, Ptolemy has shown us not only the birth of astronomy, but the birth of science as such. The discovery of the "underneath" appears to be the only revolutionary moment in the science since, as Ptolemy says, "all other things follow consequently upon it"; but an ancient revolution is practically no revolution at all.

In the rest of the chapter Ptolemy outlines many of the thoughts which might have occurred to more primitive people about the motion of the heavens, and he is careful to answer each objection in terms available to such people, as if he were trying to think through how the more advanced notions might have arisen. In passing, we might mention one of his speculations which may seem a bit odd: some ancients thought that we lived in the bottom of a great bowl. Perhaps some ancient spent the night camping out and had settled down for the evening in the bottom of some lonely crater, when a friend happened along bearing a torch; the glow of its light, as it began to creep over the round edge of the crater, might have struck him as the true image of a sunrise. Eventually someone was bound to realize that the shadows were all mixed up, since the light would have fallen first on the opposite side of the bowl; but for the moment, at least, a true act of seeing had taken place.

Reading the third paragraph of Chapter 3, one can see that the arguments slowly shift to a more sophisticated point of view. Part of this is revealed in the fact that Ptolemy begins with what one man can see from where he lives, and only later does he bring in the arguments which compare what one man sees in one place with what another man sees in another place. Only much later in the chapter does he bring up such technical considerations as refraction, the use of instruments, and, finally, philosophical arguments concerning the nature of matter itself.

The seventh chapter is a crucial one, for it is there that Ptolemy argues for the immovability of the earth. After restating Aristotle's main arguments, he presents an argument in the form of a dialogue which we shall have to consider with some care. Prior to the dialogue, Ptolemy admits that, with regard to the daily rotation, the hypothesis of the motion of the earth would be simpler; but he openly rejects it on the grounds of "what happens around us," whatever that may mean. The dialogue begins with the most forceful argument for the immovability of the earth: it would be "unnatural" for us to conceive of such a heavy body moving at such speeds; moreover, such motion would leave clouds in the atmosphere behind, there would be a constant wind, and all things would go rushing about. In other words, we would be living in a Heraclitean world. Ptolemy's counterpart then suggests that it would be possible for the atmosphere to go around *with* the world, in which case there would be no problem. In response, Ptolemy raises the question of bodies which are in the air. The argument is again reversed when the counterpart suggests that those bodies, too, might go round with the air. Ptolemy then raises the problem of how projectiles and flying bodies could be seen to change their positions relative to other bodies. Ptolemy drops the argument at that point; but since he is the expert in knowing that bodies can move with many motions at the same time, one is left with some doubts as to whether Ptolemy was not aware of the fact that the tables could easily have been turned again had the conversation been allowed to continue.

After giving a most general description of his model of the world, Ptolemy completes Book I with certain purely mathematical considerations which need not concern us for the purposes of the present article; and the same applies to Book II.

In the third chapter of Book III, Ptolemy announces the fundamental principle of his science. The passage reads as follows:

> Since the next thing is to explain the apparent irregularity of the sun, it is first necessary to assume in general that the motions of the planets in their direction contrary to the movement of the heavens are all regular and circular by nature, like the movement of the universe in the other direction. That is, the straight lines, conceived as revolving the stars, or their circles, cut off in equal times, on absolutely all circumferences, equal angles at the centers of each, and their apparent irregularities result from the position and arrangement of the circles on their spheres through which they produce these motions; but no departure from their unchangeableness has really occurred in their eternal way with regard to any disorder which might appear in the phenomena.

In the first two chapters Ptolemy had established the regularity of the year, and derived its length in days, on the basis of his own as well as more ancient observations. He had checked the length of the year from spring equinox back to spring equinox, and from fall equinox back to fall equinox.

If we had been studious enough to compare the length of time from spring equinox to fall equinox, with the time from fall to spring, we, like him, would have been worried by the inequalities which we would have encountered; but in Ptolemy's exposition the problem is not openly presented to us until a solution can be offered. This would seem to be part of Ptolemy's courtesy with regard to our natural sensitivity. He had, of course, mentioned other irregularities earlier, such as the irregularity of the planets, but those were irregularities with which we, mere common men, were already familiar. It goes without saying that, when Ptolemy lays down the goals of his science, that is to say, the conditions under which an irregular appearance would have been reduced to intelligibility, he makes no mention of the equant. Though his words at this point have been interpreted by some to include that possibility, the goals are certainly made to appear as simple as possible to the beginner.

Here the difficulties, which were not made explicit when Ptolemy chose the middle path between physics and theology, arise. There might be more than one mathematical hypothesis capable of explaining the motions of the heavenly bodies. The middle way, though it provides a more solid and precise grasp of the universe, one which establishes a foundation for the noble life, is no more of a way toward truth than the other two.[4*]

---

4 A difficulty in several English translations may cause the reader some confusion on this point. While Ptolemy often praises men by calling them truth-loving (*alethes*) he never speaks of the "true" position of a heavenly body, but only of its "precise" (*axribes*) position. Unfortunately some English versions use "true" as the translation for both words.

Ptolemy's introduction to his account of the moon, given in Chapter 1 of Book IV, is in sharp contrast to his account of the sun. As we remember, Chapter 3 of Book III began: "Since the next thing is to explain the apparent irregularity of the sun...." Ptolemy had made no mention of any irregularity prior to that statement, and the two theories—the epicyclical and the eccentric—are both laid out in full detail before we are presented with the complete account of the difficulties concerning the motion of the sun. In Book IV he begins by presenting us with a full account of the kinds of problems we will meet in connection with the moon before the solutions are given. We are allowed to fret and worry, and perhaps to think through certain problems in a way in which we were not invited to do earlier. In this way Ptolemy not only allows us to see the problems more fully, but he allows us to spend some time seeing the natural (but false) starts made by some of his predecessors before he presents his own solution. Of course, as he himself points out, the data have been sorted out by him and presented in such a way that we can almost grasp the solution before he presents it. Although the real art, that is the art of sorting the data, is hidden from our view, we can almost begin to participate in the Ptolemaic endeavor. This double-edged way of teaching will become rather critical in Ptolemy's exposition of the outer planets.

Ptolemy's account of the planets begins in Book IX with the statement of the order of the planets, and accepting the order given by the ancient astronomers, according to which Venus and Mercury are below the sun, and Mars, Jupiter, and Saturn are above the sun. Within the geocentric framework there exists no scientific way to determine the order of the planets. Ptolemy points out that the more modern astronomers place Mercury and Venus above the sun but that he follows the way of the more ancient astronomers, according to whom Venus and Mercury are below the sun. But this reverence for antiquity is strange from the mouth of the man who claims to be working within a progressive science. Curiously enough, however, one may note that the order presented, considering the transposition of the sun and the earth, turns out to be the very order which would be required by the heliocentric theory.

Ptolemy's description of what Hipparchus had done and had left undone in his treatment of the planetary motion not only seems to be a clear account of the way in which the sciences progress, but also gives us some indication as to the things which Ptolemy praises and blames. Hipparchus is praised as being zealous after the truth precisely because he refrained from offering an account of the heavens which would not be fully adequate to the task. The earlier astronomers (no doubt he has in mind people like Eudoxus who discovered the *hippopede*[5]) were satisfied with giving an account of the kinds of motion one perceived in the heavens, even if this account did not agree precisely with the phenomena. Eudoxus' theory accounted for the retrogradation of the planets in terms of strictly homocentric spheres. From the point of view of Aristotelian science, in the most formal sense of the word, Eudoxus has accounted for the kind of thing we are likely to see. Ptolemy's insistence upon precision led him to break with Aristotle on the most fundamental question posed in *De Caelo*. In *De Caelo* Aristotle derives the three possible kinds of natural motion—up, down and around—from the fact that motion may either be with regard to the equal, the greater, or the less. Bodies which move so that the distance from the center remains equal go around the center. Bodies which move so that the distance becomes less go to the center. They are the heavy bodies such as the earth. Bodies which move so that the distance becomes greater are the light bodies. When Ptolemy introduces the epicycle he tacitly rejects the possibility of any a *priori* account of the world in favor of a more precise account of the world. Since more than one hypothesis can account for the appearances, the desire for truth can best manifest itself in the desire for accuracy.

The account of the inner planets is straightforward. They wander back and forth around the sun, hence they can be understood to move on an epicycle whose center moves according to the mean position of the sun. Since there is an apparent change in the radius of the epicycle, we must assume that the deferent is not concentric with the earth. Further investigation reveals that the center of the actual path of the planet is not the same as the center around which equal arcs are made in equal times, but in fact the center of the path is actually midway between it and the earth.

The account of the outer planets does not proceed quite so smoothly. From the data, it is relatively easy to notice the peculiar fact that the cycles of anomaly plus the number of revolutions of the star in any time period is equal to the number of solar years. In order to begin his investigation of the outer planets Ptolemy must be able to isolate the several anomalies and account for them one by one. He tacitly assumes that if the position of the planet is sighted when the planet is in opposition to the mean position of the sun the effect of the epicycle will be nullified. In the data which he presents, there seems to be ample ground for assuming that if this happened to be true once it will always be true; but on the other hand there is no indication as to how Ptolemy knew it was ever true. From the appearances he knows that it is only roughly true; and

5 "Hippopede" (lit. horse-foot) is a shackle having a figure-eight shape. Eudoxus' hippopede is a figure-eight curve inscribed on a sphere; another hippopede, ascribed to Proclus, is inscribed on a plane.

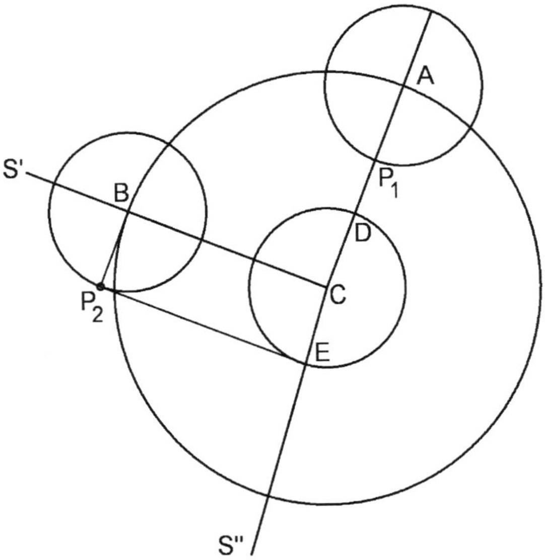

Figure 1

his leap is curious in light of his love for precision. The truth of the proposition would, of course, be obvious and even trivial on the heliocentric hypothesis.

Book XII is devoted to an account of the theory of retrogradation. At this point he chooses to mention the history of development of epicycles. Epicycles were originally intended to account for a manifest inequality obvious to all: the planets sometimes stand still in their path (as measured by the background of the fixed stars), and sometimes travel in a direction contrary to their normal behavior. Ptolemy states in general the conditions under which this appearance can occur, given the hypothesis as outlined in the previous books, and then adds the following comment:

> If the anomaly relative to the sun is taken care of by the hypothesis of eccentricity (this can be done only in the case of the three stars capable of any elongation from the sun), with the eccentric center carried eastward about the ecliptic center...

There are several peculiarities in this paragraph. First, given the fact that Ptolemy was so meticulous in the earlier books about showing the equivalence of the epicyclical and eccentric hypotheses, it is very peculiar that at no point does he give any demonstration that the theory of the moving eccentric would account for the phenomena. Stranger still is the fact that he waited until the end of his account to reveal his second hypothesis instead of presenting it back in Book IX along with his general account of the planets. But the most peculiar part of all lies in his statement that the hypothesis of the moving eccentric can only be applied in the case of the outer planets ("the three stars capable of any elongation from the sun"). Ptolemy was too good a craftsman not to be aware of the fact that it could be easily adapted to fit the case of the inner planets as well.

◆

If in Fig. 1 C be the earth, P_1 and P_2 two positions of the planet, and D and E be the two positions of the center of the moving eccentric, in either case the motion of the moving eccentric will be equal to the sum or difference of the star's speed and the speed of the epicycle. The only real distinction between the two cases is whether the mean sun happens to follow the line CB or the line CD. The interesting thing to note is that in these later passages Ptolemy brings up the case which could direct us towards heliocentricity while denying the case which would tend to lead us in the other direction.

From the works of Archimedes and Plutarch it is evident that several of the ancients had the concept of heliocentricity. It is also evident from the works of Plutarch and Seneca that the teaching of such a doctrine was politically dangerous. We, however, do not wish to stress that, since other, much more important matters, may be at stake. Schiaparelli[6] is perfectly right in saying that the reason no ancient evolved the Tychonian hypothesis is that once one sees that all of the planets can be understood to be rotating about the mean position of the sun, the transition to heliocentricity is, mathematically speaking, a trivial matter for any eleven-year-old schoolboy. Although it is true that Ptolemy's moving eccentric is not fully worked out in terms of the equant, Ptolemy could certainly have done so even though the transition is a bit more complicated than it appears at first glance. The main question is whether Ptolemy's reticence concerning the hypothesis of the moving eccentric is intended to turn our eyes away from, and thereby to turn our eyes toward, such a possibility—and, if so, why?

Ptolemy began his work by pointing out the distinction which the philosophers make between theory and practice. This distinction turned out to imply the distinction between the progressive and the repetitive sciences. One way of reading this distinction is to say that man, though he can arrive at and formulate ever more adequate and ever more intricate sciences, does not necessarily progress politically or morally.

Today, when one considers both the blessings and the dangers which confront us because of the possibilities opened up by modern science, it is almost

6 Giovanni Virginio Schiaparelli's writings have been brought together in *Le opere di G. V. Schiaparelli,* 11 vols. (Milan, 1929–1943); and *Scritti sulla storia della astronomia antica,* ed. Luigi Gabba, 3 vols. (Bologna, 1925–1927).

old-hat to wonder whether it was wise to undertake the modern endeavor. If human nature is such that modern medicine, washing machines, and airplanes—fruits of the spirit of invention—necessarily bring along with them the invention of more potent weapons of war, one is forced to raise the question that Plato may have indicated in the *Critias*: is the scientific undertaking a wise one if the political and ethical life of man cannot thereby be improved? If this question is implicit in the beginning of Ptolemy's thought, one might anachronistically pose Ptolemy's problem by asking whether it is possible to foster the spirit of inquiry without also fostering the spirit of invention. To pose such a question would mean to give up a great deal; yet, considering the threats of atomic warfare posed by the modern world, may not one be allowed to investigate such a possibility?

In the history of science it is commonplace to speak of the Copernican Revolution. Let us not ask ourselves *why* it took place but rather the more simple question: *what difference does it make*? Why should the Copernican Revolution be singled out as the transition to modern science?

To begin, as Ptolemy does, with the motion of the sun, not adding an epicycle until two chapters later, means that the pre-scientific experience of the common man forms an adequate foundation for the scientific, no matter how that might be modified and qualified by later investigations. Ptolemy's patient discussion in the first chapters concerning the most primitive views of the world, and the primitive insights which led beyond those views, greatly helps the reader in the transition from the pre-scientific view of the world to Ptolemy's science. Conversely, if one looks forward from Ptolemy, there is no proof or table in the *Almagest* which does not find an obvious counterpart in the heliocentric universe. If one takes Book XII as pointing to such an hypothesis, none of the earlier work will have been in vain.

For Copernicus, and the whole of modern science in general, there is a gulf between the pre-scientific point of view and the scientific point of view. Old wives' tales no longer serve as a foundation; the new foundations are supplied by science itself. Consider the extent to which the discoveries of Cantor, in a watered-down form, have been used in some of the less successful presentations of the new mathematics, even to first graders. If any attempt is made to popularize science, it is done by putting, in deceptively simple words, insights which arose because sophisticated questions had already been raised.

Ptolemy's solution to the problem raised at the beginning of his book seems to be that science may go where it will, so long as it does not set itself up as a radically different authority independent of the whole of the human endeavor. In defense of Copernicus, it might be suggested that, in his eyes, the content of the natural view of the world had become petrified into the scholastic view and so could no longer serve as a true foundation.

Supposing for a moment that this admittedly bizarre interpretation of Ptolemy's intention is correct, can we at best be thought of as having done anything more than making a feeble attempt to lock the barn door five hundred years after the horse was stolen? And yet it may be that an attempt on the part of both scientists and laymen to understand the truly human roots which gave rise to science—to see the problems visible within the old horizon that gave birth to the new horizon—may lead to greater unity between the practical and the theoretical sciences. I have in mind, for example, the difference between Einstein and Eddington as expositors of the new science. Eddington is quite masterful in making one at home within the new world, but one is left in awe, never knowing how he arrived there. Einstein, on the other hand, is always fully conscious of the problems which could have been apparent to men before him, and he lets us see them; thus the scientist no longer becomes a master of illusions and hence a master of our minds. In other words, the solution may be to continue the work begun by Ptolemy in his first chapters. And even if this is no solution, no way out of the modern entanglement, it may still be the case, as was once said, that "Sometimes it helps to be able to answer the question, 'what was it that just hit me?'"

Epilogue: Recollection

Some time ago, I happened to be visiting New Orleans. My hotel was across a wide and bustling street from the old French Quarter, where the streets were narrow and the buildings tall and finely worked. I was impressed not so much with what my eyes saw, but with that which can only be seen in that strange world which leaves hazy the distinction between sight and imagination. I could almost see it as it must have been: a tight little center of urban life kept small and compact in the middle of a thousand miles of nothingness. Only the sight of the *was* in the *is* can make the *is to be* what it is.

The following afternoon I wandered over to the aquarium. The penguins, so austere and august on land, yet so playful in the water, attracted my eye. As some of you may remember, they were all the rage in my childhood, and held the same fascination for children then that dinosaurs do in our own day. I suppose the last vestige of that craze is Willie the Penguin. I suddenly realized that it hadn't been too long since Scott and Amundsen had raced to the South Pole, and that must have stirred people's interest. It further occurred to me that a few years before I was born, someone must have asked a friend, "Tell me, what does one of those penguins look like?" I imagined that the answer must have been, "I'm not quite sure how to describe it; I guess it looks like a sort of queer duck." An empty phrase which I had so often heard and so often thoughtlessly used, revealed itself to be the product of human thought and human insight. Man, they say, invented the wheel. Well, *man* may indeed have slowly learned to roll a great object on a log, but it must have taken the thought of *a single human mind* to put it on an axle.

A human mind—this "I" who became interested in biblical studies, is far from the famous Cartesian *tabula rasa*. It is more like an ancient and sprawling mansion, haunted by the ghosts of thoughts that once lived and thrived in minds not my own. When I was in New Orleans, I believed I caught one of those minor and petty ghosts, the *queer duck*, and was able to bring him back to life by seeing him as part of a living human thought.

Most of us today, as we say, can feel *inertia* in the pit of our stomachs, so deep have the thoughts of one Sir Isaac Newton worked their way into our viscera. In all manner of contexts we speak of men and women we know as having or not having inertia. We think of inertia as if it were out there like an apple tree or a mountain, whereas, in fact, mountains and the apple trees are much older than the idea and even the feeling of inertia. This example we have given, the talk about the inertia that some men and woman are described as having, turns out to be a particularly fruitful example precisely because the notion of inertia which it implicitly contains is quite far from what Newton intended. It is, in fact, a kind of melange of Newton and the earlier concept of *impetus* which he had intended to replace.

Images, metaphors and language itself are the haven of the ghosts, and yet without them we could not even begin to think and would be mute as stone. The thoughts of other men from other times, men like Newton and Galileo, even going back to Phillopponus and perhaps others, seem like seeds or viruses that have buried themselves deep in our innards. These ghosts of once-living thoughts can become airborne and flit from mind to mind, and like a virus or a cancer, they masquerade as being our own.

How then shall I understand this "I" which became interested in Biblical studies, this "I" that contains the ghost of the thought of inertia, this "I" that is partly to be found between the covers of a motheaten copy of Newton unopened for centuries? Spelling myself out to myself, then, requires catching sight of these thoughts as they once lived in the minds of men long since dead.

Einstein saw this problem as clearly as any thinker ever had. In many of his writings, he often tries to retrace his thought in order to recapture the grounds upon which things first came to light. Answers are fine things; we spend our lives searching for I them. Yet, Einstein was aware of their seductive character. For him answers run the danger of settling and therefore of erasing the very questions which gave rise to them, and in so doing they erase thethemselves as answers.And in such a way as this, they degenerate into common opinion and become no more than simple prejudices or dogmas. The proofs themselves, of course do not run away. They can be written down, perhaps in some cases even simplified, and copied out into any fine textbook, but without reading, say, his paper of 1905, we do not see the problems which were at hand, available to him from within his horizon and which gave raise to his concern.

Galileo was once almost killed for teaching that the Earth moved around the sun. It disturbed those who could see the sun rise and set every day with their own eyes. Within a century everyone believed that Copernicus was right, but how many of them had reasons as cogent as those who had once claimed, "The sun rises every morning; I see it with my own eyes." Surely in grade school I was taught why it might appear that the sun had moved even though in fact it was I who had moved, but since the explanation came from someone who already believed that the Earth moved, I never saw the problems that would cause one to leave the more obvious way of understanding the appearances. I have great admiration for my fourth grade teacher, but I seriously doubt whether she knew anything more than that today "we all know" that the earth moves.

There is another side of things. As a child, I had a rather strange view of the world in which I lived. Of course I somehow knew that Shakespeare wrote in English, and that he had written a very long time ago. At the same time, I was confused by the fact that the grandfather of the boy next door could speak English as well as I could. "Grandfathers speak Yiddish," so I thought.

I was raised in the reformest of the reform. No hats allowed in the temple, and all Hebrew was sung to the tunes of Lutheran hymns. But the holidays were different; then we would visit my grandfather's synagogue. There the tunes were somehow better fitted to the words. They seemed to be more within their own horizon. Even as a child, I had the sense that this thing we call the "I" flowed back into a barely sensed world, partly into Grandpa's shul, and, as I later realized, partly back into the pages of Newton's *Principia*.

Little boys grow up, and I turned from grandpa to Newton and Plato. My parents had wanted me to attend Brandeis University, but fortunately for me, Brandeis had no room in those days for a young man with cerebral palsy; so there was no family squabble. My own imagination had been caught by a man named Jasha Klein of St. John's College. Klein was fascinated by the act of trying to see how the everyday world revealed itself to each person and in what way each particular horizon implied, and hence led to, a world beyond itself.

For Klein this facing the world meant coming to terms with those books and thinkers of the past which have helped to form us and our ways of thought—not because we had read them, but precisely because we had *not* read them but had only absorbed them passively and thoughtlessly, through cheap talk and terms of speech we supposed ourselves to command.

To understand ourselves, our own haunted houses, then, means to see that horizon opening up by reconsidering the labor of that thinker for whom it first opened, and to listen his own voice. Knowledge, for Klein, was not something that could be written down in a book; knowledge demanded our own voice.

Shortly before I graduated from St. John's, Mr. Klein introduced me to Professor Leo Strauss. To my good fortune, Strauss was to spend the following year in Israel, which meant that I did not have to make the hard choice between going out to Chicago, or going first to Israel in order to learn Hebrew. It also meant that Strauss would be spending a good bit of his time on Judaic studies. He was most generous with his time; and we would spend many hours together each Shabbat reading *Bereshit* (Genesis). At five or five-thirty there would always be tea and little cakes, and on rare occasions we would just relax with the TV show "Gunsmoke." He was particularly attracted to the delicate character of Miss Kitty.

For Strauss, understanding ourselves primarily meant understanding what for him were the two fundamental alternatives that uneasily sat together to form the foundation of our western world: the Bible and Greek philosophy. From Strauss I learned to respect the Bible enough to expect it to hold its own in any conversation. For me, piety meant having the expectation that the Bible, if pressed, would seriously address even my most impious questions.

Robert D. Sacks, A Biographical Sketch

Robert D. Sacks was born in Akron, Ohio, in 1931; and in units of Bar Mitzvah (where one Bar Mitzvah = 13 years), he now claims to be "seven Bar Mitzvahs old." This is perhaps an unorthodox unit of time, yet singularly appropriate for the "unorthodox" Dr. Sacks. In the Jewish denominational sense, Sacks is both unorthodox (Reform) and Orthodox; and such dichotomies abound in his life story. His childhood centered on a family that was quite liberal in its Jewish practice; he recalls worship services which incorporated Lutheran hymnal music for the Jewish liturgy. Yet his grandfather, who clearly had a strong influence on Sacks, spoke Yiddish and followed a more traditional Judaism. While Sacks still considers himself Reform, he describes himself as a Reform Jew who feels most comfortable with Orthodox Jews. This attitude is typical of Sacks, who clearly enjoys being "out of place." Were he to live in Israel, he maintains that he would probably associate closely with the Druze, a breakaway sect derived originally from Islam. And as an ancient Greek, it wouldn't be Plato or Socrates whom Sacks would be following but Apollonius, the great geometer, best known for his work on conic sections.

Robert Sacks began talking at the age of two, but he learned much later that he had been trying to communicate verbally without great success for a full year before his father realized that he was speaking rather than just making random sounds. Such is the nature of the hurdles that Sacks has needed to climb as a result of his cerebral palsy. At a Robin Hood-themed camp in Vermont for kids living with this condition, six-year-old Robert Sacks staked a claim on a cabin as his own and declared a sense of individuality and self.

Sacks remains intrigued by the meaning of "I" and has written extensively and eloquently on the topic. His brother Jim became the right hand colleague of Jacob Moreno, the founder of psychodrama; but Sacks' earliest aspirations were for a career as a chemist. He was fascinated by the concept of combining chemicals and forming something new. (In a way, Sacks' academic work is indeed a form of chemistry, with words and ideas rather than chemical ingredients). Cerebral palsy prevented Sacks from attending Brandeis University, the school of choice of his parents. This presented little hardship for him, however, as he was already more interested in attending St. John's College.

Jasha (Jacob) Klein, who served as dean of St. John's during Sacks' college years, not only captured his imagination before he matriculated, Klein also later introduced him to another mentor, Klein's fellow émigré Leo Strauss. (Whether by choice or happenstance, Sacks has gravitated toward these brilliant Jewish intellectual figures.) From a conversation with Klein when Sacks was a mere 18 years old, Sacks suddenly felt that the world made sense. From Strauss, Sacks learned how to read; from Klein he learned why to read. Klein, whose influence on St. John's College was enormous, taught Sacks the difference between a good book and a great book; and he planted in Sacks' mind the image of books as "ghosts of other people's thoughts."

Sacks planned to attend the University of Chicago after receiving his undergraduate degree in order to learn from Professor Strauss after first spending a year in Israel to study Hebrew. Fortuitously, Strauss also spent that year in Israel, so Sacks began his mastery of Hebrew and his studies with Strauss simultaneously rather than sequentially. Strauss was a hospitable though demanding host at their weekly afternoon sessions. Sacks recalls Strauss saying at an early Shabbat get-together, "Mr. Sacks, you don't know anything about the Bible. Please come again next Shabbat."

Obviously Sacks eventually passed muster, receiving his Ph.D. from The Johns Hopkins University in 1961 with stints at the Ecole des Langues Orientales Vivantes in Paris; Hebrew University, Jerusalem; and the University of Chicago. Sacks enjoys recounting that his time in Paris was dominated less by academic pursuits than by wandering the city, haunting bookstores and cafes, and getting lost. To this day, he happily points to a picture, on his wall, of Notre Dame Cathedral with the bookstore in the foreground. He claims, with some evident modesty, not to have spoken French when he went to Paris; but his skill with languages has enabled much of his subsequent career. His biblical studies required some knowledge of Aramaic; and though he doesn't consider himself a Talmud scholar, his mind clearly works with an underpinning of Talmudic logic.

At St. John's, Sacks acquired his first knowledge of Greek, and it was there that he became engrossed in the work of Apollonius. Apollonius, whom he greatly admires, taught him that the point of reading a book is not to find out the answer but to find out what question to start with (note: not unlike the Talmud). Sacks himself says of Apollonius (page 158), "I shall leave you to follow his backwards journey to see how he managed to arrive at his beginning." Perhaps, then, this is an apt end (or beginning) of our biographical sketch of Robert D. Sacks.

Rabbi Jack Shlachter
Congregation Hamakom, Santa Fe, and
Los Alamos Jewish Center, Los Alamos

Index

Y

Z

Made in the USA
Monee, IL
14 November 2026

68482486R00118